Water Justice

What You Don't Know Could Kill You

Susan Blacklin

BLACKLIN
PRESS

WATER JUSTICE

WHAT YOU DON'T KNOW COULD KILL YOU

FROM THE AUTHOR OF WATER CONFIDENTIAL: WITNESSING JUSTICE DENIED

SUSAN BLACKLIN

BLACKLIN
PRESS

Blacklin Press, British Columbia, Canada

www.susanblacklin.com

Editor: Christina Myers

Cover and interior design: SGA Books

Cover images via iStock: Adaask, Selensergen, PowerofForever

Ebook ISBN-13: 978-1-7774654-2-1
Paperback ISBN-13: 978-1-777-4654-3-8

For the owners of Canada and those we elect to represent us.

Author Note

I am a messenger sharing many stories and situations, which were brought to my attention.

All opinions expressed in this book are mine, based on my personal experiences and research.

Others may form a different opinion.

"You write in order to change the world."
James Baldwin

Also by Susan Blacklin

Water Confidential: Witnessing Justice Denied. The Fight for Safe Drinking Water in Indigenous and Rural Communities in Canada

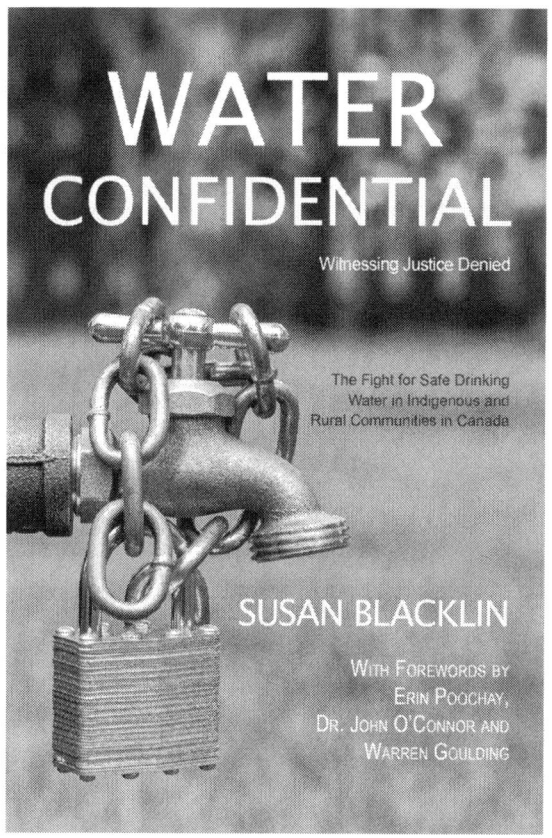

Available from Caitlin Press: https://caitlinpress.com/Books/W/Water-Confidential

Glossary of Acronyms

AANDC: Aboriginal Affairs and Northern Development Canada
AER: Alberta Energy Regulator
ACFN: Athabasca Chipewyan First Nation
AFFF: Aqueous film-forming foam
AI: Artificial intelligence
AFN: Assembly of First Nations
BCAFN: British Columbia Assembly of First Nations
BCNDP: British Columbia New Democratic Party
BWA: Boil water advisory
CAAF: Climate Action and Awareness Fund
CBC: Canadian Broadcasting Corporation
CDC: Centres for Disease Control and Prevention
CEU: Continuing Education Unit
CI: Charity Intelligence
CNL: Canadian Nuclear Laboratories
CRA: Canada Revenue Agency
CSIR: Centre for Sustainable Infrastructure Research
CWA: Canada Water Agency
CWN: Canada Water Network
CWWA: Canadian Water and Wastewater Association

DGR: Deep Geological Repository

DNC: Do not consume advisory

DWA: Drinking water advisory

ECCC: Environment and Climate Change Canada

EFSA: European Food Safety Authority

EPA: Environmental Protection Agency (USA)

EVR: Elk Valley Resources Operations Limited

FCM: Federation of Canadian Municipalities

FEIs: Freshwater Ecosystem Initiatives

FNHA: First Nations Health Authority

FOI: Freedom of Information

GM: Genetically modified

GMO: Genetically modified organism

IBROM: Integrated Biological Reverse Osmosis Membrane (treatment system)

ISC: Indigenous Services Canada

KOW: Keepers of the Water

LNG: Liquefied natural gas

LPD: Litres per day

MAC: Maximum acceptable concentration

MLA: Member of the Legislative Assembly

MOU: Memorandum of understanding

MP: Member of Parliament

MSW: Master of Social Work

NDP: New Democratic Party

NGO: Non-governmental organization

NPO: nonprofit organization

NPRI: National Pollutant Release Inventory

NRRM: Northern Rockies Regional Municipality

NSERC: Natural Sciences and Engineering Research Council of Canada

NRC: National Research Council

NWMO: Nuclear Waste Management Organization

PAC: Political Action Committee

PCN: Prevent Cancer Now

PFAS: Per- and polyfluoroalkyl substances
RCMP: Royal Canadian Mounted Police
RO: Reverse osmosis
RR: Roundup Ready
SDWF: Safe Drinking Water Foundation
SFNWA: The Saskatchewan First Nations Water Association
SMR: Small modular reactors
SRC: Saskatchewan Research Council
SSHRC: Social Sciences and Humanities Research Council of Canada
SUMA: Saskatchewan Urban Municipalities Association
TFEC: Tax-Filer Empowerment Canada
THMs: Trihalomethanes
UBCM: Union of British Columbia Municipalities
UNDRIP: United Nations Declaration on the Rights of Indigenous
Peoples
URL: Underground Research Laboratory
WHO: World Health Organization
WLON: Wabigoon Lake Ojibway Nation
WSBC: WorkSafe British Columbia
WTO: Water treatment operator

Foreword by Ash Smith,
Founder, Windrush Against
Sewage Pollution (UK)

When I first heard from Susan Blacklin, calling from Canada, I was expecting something entirely different. Being a campaigner against sewage pollution, I thought she was going to offer me some sympathy for the water industry scandal that has unraveled in the UK and a smug account of how it should be done from the home of the salmon run and the land of pristine rivers, crystal clear meltwater and natural beauty.

Well, that didn't happen and I was genuinely shocked and disappointed to hear that we had so much in common. With almost ten million square kilometers of Canadian land home to about forty million people versus about 244,000 square kilometres creaking under around sixty-nine million thirsty people here in the UK, that really surprised me.

My theory that we can afford to behave pretty stupidly as long as there are not too many of us doing it per hectare, has probably been eroded by the strength and enduring nature of the pollutants that our increased chemical sophistication, for want of a much better word, has brought with it. Sewage pollution is no longer water mixed with human waste products and a bit of soap and bleach, as a trip round your local supermarket's washing and cleaning aisles will confirm; read the labels and shudder at what we are lining up ultimately for our own consumption. The list of drugs, hormones, chemicals, bacteria and viruses and their enduring nature is still growing.

Susan addresses this and much more with a comprehensive flight over Canada's water, exploring the complex and insidious aspects of regulatory capture and the commercial incentives that shape corporate behaviour across the globe. Her mixture of real life stories and the fruits of meticulous research lay bare a reality that ought not only to inform and engage the reader but, more pressingly, to be a wake-up call, not only for government, but for anyone with the imagination to realise that water pollution, especially at aquifer level, is very easy to do but incredibly difficult to fix. And then what are you going to do? Drink beer and shower in it? I appreciate that will excite a few people but you get the idea.

The asbestos pipe risks that Susan reveals are worth nightmares of their own and the complexities of governance, responsibility and regulation, or rather the failure of all of these to ring a very loud bell to wake up a population heading towards a cliff edge, brings a chilling message. Avoiding this disaster is probably going to be down to the victims: the public who need to start making a fuss and a big one.

And thus the messages this book brings are so important and with the knowledge it delivers should come inspiration and action. It is going to have to, because the tipping point of stupidity/hectare makes no exceptions.

I was left with the feeling that Canada is edging towards the position that the UK was ushered into in 1989 by ten years of ruthless underfunding and poor governance harming previous decades of rebuilding infrastructure after the "Great" Wars. Later on, as prosperity increased, there came other priorities that were more important than looking after our water supply and where we live, or so we thought.

The solution was made to look complex and expensive, way beyond what a government could achieve with so many competing demands. The Welsh and English components of the UK fell for the promises that privatisation would bring investment and excellence in engineering but instead it brought cash extraction from captive customers and financial engineering that has loaded our industry with debt and almost taken us over the cliff edge.

From where I am sitting, near the shadow of a once-vibrant river in the Cotswolds, England, it seems to me that *Water Justice: What You Don't Know Could Kill You*, brings a message that should have come twenty years ago for you and forty for us, but no one would have listened back then, would they?

Ash Smith, Founder, Windrush Against Sewage Pollution

Foreword by Julian Branch, Prevent Cancer Now

I first became aware of Susan Blacklin as a result of my ongoing research into asbestos cement water pipes. She and others were organizing a petition calling for stronger regulation of Canadian water. When we connected, we discovered a central theme: a disconcerting lack of political accountability with regard to water. There isn't a politician alive who won't tell you that they are deeply concerned about safe, clean drinking water. The truth however, is something quite different.

After reading Susan's first book *Water Confidential: Witnessing Justice Denied*, it was obvious that there are gaping holes in the regulation of Canadian water. Because of my work in the area of asbestos cement water pipes I was acutely aware that asbestos is not regulated in Canadian water. It has been regulated in American water for more than three decades. The reason given in the US in 1992 was "to protect against cancer." Yet here in Canada, we continue to maintain that there is insufficient proof that swallowing asbestos is harmful. Politicians at all levels cling to that belief with determined desperation.

After five years of working with my local community association, trying to raise awareness about asbestos cement water pipes I have been told it told it "does not want to ruffle the feathers" of city hall because that's where the association gets its funding for hockey programs. The community association went as far as to remove an information infographic on asbestos cement water pipes from its social media site. When my name appeared in a national story recently about the dangers of swallowing asbestos, the former co-president of my community association sent me a series of vile obscenity-filled emails chiding me for my work. I assured him that he is free to drink as much asbestos as he wishes. I also assured him that I will continue to raise awareness of this important health issue. My local city councillor has refused to hold an information meeting to raise awareness of the issue, as has my local MLA. My MP conducted his own research, and promised to go to Ottawa, and ask tough questions. That was three years ago. I am still waiting for action. Very recently I appeared before my city council to raise the issue of asbestos cement water pipes. I was referred to as "insufferable" by a former city councillor for daring to talk about the pipes.

Asbestos cement water pipes are a perfect issue for politicians. Out of sight. Out of mind. Very few people are even aware that Canada still has almost 14,000 kilometres of old asbestos cement water pipes. When politicians first hear of this issue, they are aghast, and promise action. However, after calculating the cost involved to replace the pipes, and the political return, the line goes dead.

What Susan does with her dogged pursuit of the truth related to a glaring lack of regulation of Canadian water is to shine a light on an incredibly important issue that few of us even think about on a daily basis, as we twist the tap to fill up a glass. I certainly took it for granted until I discovered that my city of Regina, Saskatchewan has 600 kilometres of old asbestos cement water pipes.

Susan is to be commended for the painstaking, rigorous and meticulous work she has done on putting the sorry state of Canada's drinking water regulations, and the people overseeing that system, under a microscope. Canada needs more Susan Blacklins.

Julian Branch, Board Member, Prevent Cancer Now

Introduction

Many readers of my memoir, *Water Confidential: Witnessing Justice Denied*, were astonished to learn about how my late ex-husband and I worked tirelessly to improve the quality of drinking water in Indigenous communities in Canada. I will refer to him throughout this book as Dr. H2O, which is what many First Nations people called him and appeared on his personalized licence plate.

Some readers of *Water Confidential* shared concerns and brought additional issues to my attention. They inspired me to write this book. *Water Justice: What You Don't Know Could Kill You* explains how rural, urban, and First Nations communities in Canada face serious health threats from unsafe drinking water. Don't think that if you live in large cities or small towns, your water is safe.

Canada is the only G7 country that does not have national drinking water regulations. Canada's G7 presidency offers a crucial test of whether the new government's vision of a strong Canada extends beyond domestic concerns. Despite water's fundamental connection to both economic stability and national security, it was notably missing from all six G7 leaders' statements. Over twenty organizations and networks across Canada have now written to Prime Minister Mark Carney, urging action on the G7 Water Coalition announced in Italy in

2024. Their unified message is clear: Canada must use its G7 residency to champion global water security.[1]

Unlike the United States and many countries in Europe, Canada does not have legally enforceable safe drinking water regulations, which causes inconsistency across its provinces and territories. Canada also has no microbiological treatment standard to safeguard against pathogens such as Escherichia coli (E. coli). The country's existing guidelines are applied unevenly. Depending on the jurisdiction, some are mandatory and others are voluntary.

I detail what and who is polluting our waters, and the ramifications of the government's failure to make safe drinking water a national priority. The lack of water standards is not due to inadequate funding. Rather, it is a lack of political will, due diligence and accountability, which has continued for decades, and is still continuing today.

I wish it were better now than during the period that I covered in *Water Confidential*, which was twenty-plus years ago. However, too many people in charge remain uninformed or uninterested in addressing the issue.

Recently, a national journalist spent hours interviewing me for an article about water quality but later called to tell me their editor did not want to publish the story. They were instructed to write about sustainable electricity instead. Apparently, our water—the essence of life—is either too easy to ignore or too alarming to address.

I once gave a presentation over Zoom to an audience composed mostly of university doctoral students in sustainable agriculture. One man from Uganda introduced himself, speaking slowly, choosing each word carefully. He said that he had scrimped and saved to complete his doctorate in soil sciences. Coming to Canada had been a dream for him. Once here, he found work with an environmental consulting company. That is when his dream was shattered.

He said, "I know corruption is prolific in Uganda, I didn't expect it to be the same here in Canada. The consulting company told me what I had to write in my assessments and reports." He revealed that he had been expected to give a pass to polluters, and thereby to pollutants. Environmental concerns were often ignored because executives of his company feared it would never receive future contracts if it did not approve polluters. He quit that job to drive a taxi and volunteer to help immigrants grow food in community gardens. He said he slept better at night than when he was covering up health-threatening pollution.

Many readers of *Water Confidential* were outraged by what they learned. One man, who bought the book at a book signing, returned an hour later to buy another copy. "I began reading it and can't put it down," he told me, adding, "I'm going to give this to my MP [Member of Parliament]; he needs to read this."

The Congress of Humanities selected *Water Confidential* to be included in the curriculum for Indigenous Studies in universities across Canada. British Columbia's Department of Education also endorsed *Water Confidential*, recommending its inclusion in high school science and social studies curricula.

I have included short, positive stories about activists who are trying their best, who are determined to make the world a better place against a tide of denial and indifference. I hope you will be motivated to take action, too!

I hope that *Water Justice* will educate and encourage you to become an advocate for safe drinking water. May this book also inspire you to ask questions of our leaders and to expect appropriate answers. We must take control of our water and our lives here in Canada. I hope this book will be your call to action!

Susan Blacklin
    Author, *Water Confidential* and *Water Justice*

Table of Contents

Part Three
Friends of Government

Part Four
Urban and Rural Communities at Risk

Part Five
Is Any Drinking Water Safe In Canada?

Part Six
Advocacy

Part One

Safe Drinking Water 101

Chapter 1

Standards for Water Safety?

Access to clean water is one of the most fundamental human rights. It helps us to avoid unnecessary exposure to many diseases that can harm our health and lead to premature death. According to the World Health Organization (WHO), safe and accessible water is crucial to public health, whether it's used for drinking, bathing, food production or recreation. Unfortunately, 844 million people worldwide lack access to safe drinking water, and adequate water for sanitation and hygiene. This includes many people who live in Canadian communities.

The WHO has established international standards for safe drinking water, stipulating that water intended for human consumption must be free from organisms and concentrations of chemical substances that may be hazardous to health. Additionally, water temperature is important (microbial safety that is aided by hot water should be balanced with burn prevention and palatability that is aided by cooler water), absence of turbidity is necessary, colour is used as a key indicator of water quality, and any strange tastes or smells can signal pollution or treatment issues. It also states that "construction, operation, and supervision of a water supply, its reservoir, and its distribution systems must exclude any possible pollution."

The greatest threat to safe drinking water, according to WHO, is

Susan Blacklin

sewage (human or animal excrement), which results in deadly *E. coli*. I am concerned that, in addition to the ongoing impact of toxins/pollution, increasing drought conditions are magnifying the serious impact of pollution. When the volume of water is reduced, the concentration and relative impact of toxins surely increase.

Other causes of contamination arise from both natural and human-made sources, such as chemically contaminated soil (e.g. arsenic or radon), leaks from landfills or underground fuel tanks, as well as pesticides, and large industrial livestock farms and wildlife.

Chapter 2

Waterborne Diseases

There are many waterborne diseases that lie dormant and do not result in symptoms for years, sometimes for decades. Most are caused by consumption or contact with contaminated water that contains viruses or bacteria. Diseases from contaminated water can also be spread while bathing, washing, and eating food prepared from a contaminated water supply.

Waterborne diseases can reveal themselves through different symptoms. While diarrhea and vomiting are the most common, symptoms such as rashes, as well as ear, respiratory and eye problems, have also been associated with waterborne diseases.

I helped a co-worker who became ill one afternoon. She went home to sleep, but two hours later, she woke up vomiting blood. When I heard her weak voice asking for help, I knew she was seriously ill. I called an ambulance. On the way to the hospital, the paramedics nearly lost her.

Her teenage daughter and I waited several hours, hoping and praying. Every hour or so, we were told, "She's not out of the woods yet; everyone is still doing their best for her." Finally, a doctor came in, wiping perspiration from her brow, and told us that she was going to make it.

They moved her to intensive care.

My friend was lucky to survive. She had been diagnosed with a waterborne illness known as *Helicobacter pylori (H. pylori)*, an infection that can last a lifetime in a host if not properly treated.[2] Most people are unaware that they have *H. pylori* and show no signs of infection. Others can develop long-lasting inflammation in the stomach lining, called chronic gastritis.[3]

H. pylori infection is one of the main causes of gastric (stomach) cancer and duodenal peptic ulcer disease. It can also lead to mucosa-associated lymphoid tissue lymphoma.

H. pylori infection is a common bacterial infection worldwide. In Canada, the prevalence of *H. pylori* infection is estimated to be between 20 and 30 percent of the population.[4] Certain groups, such as those who live in Indigenous communities in Canada, and immigrants from high-prevalence areas (such as South America, Africa and Asia), experience higher infection rates.

Wouldn't it be prudent and cost-effective to run a simple blood test on all individuals who are likely to have the *H. pylori* infection and provide antibiotics to all who test positive, thereby avoiding a life-threatening experience such as I witnessed?

Pathogenic microorganisms in drinking water, either from fecal contamination or natural aquatic environments, can cause a range of illnesses from gastrointestinal problems to severe infections, especially in vulnerable populations. Successful management requires understanding these pathogens, their transmission, health effects, and initiating effective controls to ensure safe drinking water.[5] Effective water treatment systems are crucial to avoiding waterborne diseases.

Chapter 3

Drinking Water Advisories (DWAs)

DWAs are important alerts that are issued to warn people that their water is unsafe. They are issued when microbiological contamination is suspected or confirmed, when treatment or distribution system failures occur, when chemical contamination is detected or suspected, when natural events compromise water safety, or when operational issues create uncertainty.

The three most common types of DWAs are "boil water advisory" (BWA), "do not consume advisory" (DNC) and "do not use advisory." A BWA indicates that tap water should be brought to a rolling boil for at least one minute before drinking it, cooking with it, or using it for brushing teeth, making ice, washing produce or preparing infant formula. Only water that has been brought to a rolling boil for at least one minute should be used to bathe infants, young children and immunocompromised individuals. A DNC is issued when the sampling of water indicates contaminants, such as lead, that boiling will not eliminate.

During this advisory, water certainly should not be consumed even if it is boiled. While it can be used for bathing infants, young children and people with compromised immune systems should be given sponge baths to prevent them from ingesting water. "Do not use," the most severe advisory, means water must not be used for any reason because it poses a serious threat to public health.

This system confusing. The First Nations Health Authority (FNHA) posts DWAs, (Monthly Drinking Water Advisories in First Nations Communities in BC)[6] which are based on a summary of water systems with five or more connections (community water systems), and smaller water systems that have public facilities such as campgrounds (public water systems). Public water systems include both nation-owned and privately-owned systems.

Water systems on leased land are not included. Water Today publishes the national DWAs (advisory maps, watertoday.ca) for both First Nations and settler communities, which are updated regularly.

Indigenous Services Canada (ISC) posts information on community water systems that are owned and operated by the First Nations communities and funded by ISC. The data sets used by FNHA and ISC are slightly different, but FNHA includes all those that ISC posts.

In the summer of 2025, there were thirty-two DWAs in First Nations communities, and over 1,400 in rural communities. This does not indicate the magnitude of poor water quality because long-term BWAs are only those that have been in place, consistently, for at least one year.

Many households use water jug filters, thinking they improve water quality, but Dr. H_2O once refused to endorse such a product, stating it only removes chlorine odour and does nothing to make water safe to drink.

WHISTLEBLOWER HONOURED

Dr. John O'Connor received Ryerson University's inaugural Peter Bryce Prize for Whistleblowing after exposing elevated cancer rates in Fort Chipewyan, Alberta, linked to nearby oil sands operations.

In 2006, O'Connor observed disproportionately high rates of rare and common cancers among residents. His whistleblowing triggered severe backlash; three Health Canada physicians filed four complaints against him, leading to a College of Physicians investigation. O'Connor faced significant legal and professional opposition for his claims.

Despite this resistance, a 2009 Alberta Cancer Board study vindicated him. It confirmed that cancer rates were approximately 30 percent higher than expected. The experience "changed my life forever," O'Connor said, highlighting the personal cost of challenging government denial.

Part Two

First Nations Water Challenges

Chapter 4

First Nations Clean Water Act (Bill C-61)

During my book tour for *Water Confidential,* in the spring of 2024, a middle-aged woman approached me in tears after my presentation.

"I am a pediatric nurse. I am so ashamed," she said. "I have always been annoyed at a doctor who spent more time with the (Indigenous) kids … now I know why. I can't believe how much he knew, how much he cared. I never realized that they didn't have safe drinking water."

She asked me to sign a copy of my book and promised to share it with other staff in her hospital. In December 2023, ISC Minister Patty Hajdu introduced the First Nations Clean Water Act, setting drinking water standards in First Nations communities. This legislation, also known as Bill C-61, claimed it would protect fresh water sources, create minimum national drinking water and wastewater standards, and deliver sustainable funding for maintaining water quality.[7]

The First Nations Clean Water Act was proposed to establish minimum standards for water services on First Nations lands and recognizes the inherent right of First Nations to self-government regarding water management. The bill acknowledges clean drinking water as a basic human right and a fiduciary obligation of the Crown, and addresses concerns about inadequate funding, insufficient engagement, and potential infringements on Aboriginal and treaty rights.

The bill promised to develop federal regulations to ensure access to safe, clean, reliable drinking water, and the effective treatment of wastewater on First Nations' lands. It also pledged to lay the foundation for a new First Nations-led water institution to support communities, to be known as the First Nations Water Commission.[8]

I did not view the bill as a positive move for Indigenous people. Neither did many of the leaders of First Nations communities. Unless the federal Bill could supersede all provincial laws, I could not see how it could help Indigenous people to protect their source waters. Plus, I felt the government was downloading liability instead of giving First Nations responsibility, due to the great majority of water treatment systems being ineffective and incapable of producing safe drinking water. Indigenous leaders objected to Bill C-61 because it lacked adequate funding, failed to protect source water, and undermined Aboriginal and treaty rights by including sweeping federal powers that narrowed Indigenous jurisdiction. Critics argued the bill was drafted without meaningful consultation and consent from First Nations, creating confusion and failing to incorporate their governance and water needs.

I believe that less than 10 percent of First Nations water treatment systems can produce truly safe drinking water, that is to a standard which meets or exceeds WHO recommendations. The federal government presently has jurisdiction over all First Nations water treatment systems, and it has failed miserably. If the government is serious about reconciliation, providing effective water treatment plants to all First Nations communities would be a reasonable first step.

Dr. H_2O successfully trained many Indigenous water treatment operators (WTOs). I was introduced to many Indigenous cultural celebrations and developed a deep appreciation for times when they honoured the water spirit or sang similar mesmerizing chants, asserting their respect for Mother Earth, and the vital water systems.

My hope is that their respect can be enhanced/complemented by updated water treatment systems. The WTOs whom Dr. H_2O trained felt the Integrated Biological Reverse Osmosis Membrane (IBROM) treatment system was compatible with their cultural beliefs, as Indigenous Peoples object fiercely to the odour and taste of chemicals such as chlorine, as do I. The IBROM system can be taught easily as it requires virtually zero chemicals and rarely any chlorine, saving substantial chemical costs as well as protecting the environment.

The quality of drinking water for all the communities under DWAs, and those who should be under DWAs but are not, must also be assessed and addressed. I suggest that one of the biggest problems for First Nations leaders is determining who their allies are and who they can trust, which really shouldn't surprise anyone considering how colonialism has treated them. I suggest that truly safe drinking water will only become a reality when scientists and Indigenous leaders form partnerships working together for the greater good of each community.

Protection of source waters is the critical point, and source waters presently fall under varying provincial laws. Unless Bill C-61 supersedes all provincial and/or territorial laws, First Nations community leaders will be no further ahead if they cannot ban industry from releasing toxic effluent into waterways, which is a provincial mandate.

If Bill C-61 demands regulations, why not simply apply regulations now and protect all Canadians?

Without a federal change, where does this leave rural Canadians? They are also subjected to varying, often weak, provincial standards, and are entitled to consistent, legally enforceable regulations for their drinking water quality and source water protection.

In 2020, a consortium led by Concordia University's Institute for Investigative Journalism surveyed WTOs, managers, and public works employees from 122 First Nations across Canada. The survey found that two-thirds of First Nations water operators earned less than the median wage for operators in their province. Their pay was sometimes close to minimum wage, despite being on call around the clock. Their poor wages persist because, although the federal government has long been aware of the underfunding for operation and maintenance of on-

Susan Blacklin

reserve water and wastewater systems, ISC has only recently begun to update its funding formulas from 1998.[9]

Moosomin First Nation in Saskatchewan received an IBROM water treatment system in 2013. Nathan Martell, a certified WTO, began working at the Moosomin water treatment plant in 2003. He was trained in the IBROM process by the WTOs who established their own Advanced Aboriginal Water Treatment Team to give back to others after Dr. H$_2$O had trained them.

Martell's skills were highly sought after across Saskatchewan, leading to his employment in wastewater treatment at the City of North Battleford, about forty kilometres south of Moosomin First Nation. His new job offered better pay, pension, and benefits, which his First Nation couldn't afford. Despite the lucrative position, Martell continues to work at the Moosomin water plant because there's no one else to take over if he leaves.

Martell says, "It's our community. We grew up here, and my family's from here. ... I don't want anyone to get sick or anything to go wrong. This is my home. You can't turn your back on home."[10] As far as I know, Martell continues to work two jobs to protect his home community. Nathan is not alone; there are so many Indigenous people doing their very best, in challenging situations, against all odds in a colonial system. Here are more Indigenous people striving for the greater good of their people.

PLEDGES TO ATTACK CLIMATE CHANGE

Since 2016, three state-level projects have brought reliable power to 140 million people in India, laying out more than 4,300 kilometres of power lines and rolling out smart meters and digitised billing.

What's even more impressive is that this has happened even as India has reached its target of 50 percent clean power capacity way ahead of schedule: Half of the country's 438 GW of installed electricity capacity now comes from renewables, hydro, and nuclear, fulfilling its 2030 Paris pledge five years early.

Chapter 5

Keepers of the Water (KOW)

Where we live often determines our health and life expectancy. A shameful example of this is Swan Hills, a town in northwest Alberta. In *Water Confidential*, I shared the plight of people living near the Swan Hills toxic waste dump in the late 1990s. Three decades later, the affected communities still have no adequate water treatment systems and drink polluted water.

They are not alone.

KOW, founded in 2006, is an Indigenous advocacy group in Alberta that works to protect ecosystems and combat climate change. They never stop speaking out about the need for safe drinking water for First Nations.

Early in 2025, Keepers made yet another call for support to protect Swan Hills and neighbouring House Mountain and Chrystina Lake areas from potential contamination and pollution risks from the Swan Hills Hazardous Waste Treatment Centre. Keepers challenged the Alberta government over the Swan Hills facility's ongoing pollution for over a decade. By early 2025, they were in mediation with the Alberta Environmental Appeals Board and sought to demonstrate strong community support before the deadline of February 28, 2025.

They hoped the Board would hear their concerns and take action to protect the environment.

However, the outcome was disappointing. The Keepers reported that they decided to shelve their appeal. Their eco-justice lawyers recommended strategy is to pursue funding for additional environmental and health testing, then wait until the facility's decommissioning plan is submitted, to raise their objections.

Their problems are getting worse.

On May 26, 2025, residents of Swan Hills were ordered to evacuate as powerful, unpredictable winds fanned the flames of a nearby wildfire. Residents were told to gather pets, important documents, and have enough food, water, fuel, and supplies to last at least three days.

I question what will happen if the toxic waste starts burning? Maybe the question should be *when*, not *if*. We are in unprecedented times.

Forest fire smoke poses a health risk due to chemicals and their reaction to heat. Some areas of the forests around Swan Hills contain low levels of PCBs, which are man-made industrial chemicals that were widely used for their resistance to heat and electrical insulating properties. In a forest fire, the PCBs will volatize in the air and be present in the smoke. The PCBs may also convert to other toxic chemicals. [11]

"The smoke from the Swan Hills (Edith Lake) fire is more dangerous even than 'regular' forest fire smoke. That's because all the emissions from the Swan Hills Hazardous Waste Treatment Centre are concentrated in the surrounding forest. These dreadful contaminants are going back into the air in the smoke from the current wildfire."

Storing toxic waste on top of a mountain in an area that has a high probability of burning at some point shows a blatant disregard for people's health. Toxicity is threatening the lives of people drinking the water downstream.

Please don't think Swan Hills does not concern you. A map shows the raging fires, but it does not show where the wind had already spread the toxic air. Cumulative effects are not being considered. The media are covering fires burning in Manitoba and Saskatchewan, yet nobody seems to be making the public aware of the Swan Hills debacle.[12]

THE NECESSITY OF WATER

Seth Maxwell, founder and CEO of Thirst Project, the world's largest youth-led water activism organization, stated he believed the global water crisis was "the single-most pressing humanitarian crisis" facing the global community.

He explained that every cause was impacted first by water. Regarding education, he noted schools couldn't function if children were sick with waterborne diseases or spending six to eight hours daily collecting contaminated water: "If you care about education, you care about water."

Similarly, he argued agricultural development and combating food insecurity were impossible without water: "If you care about hunger, you care about water." Maxwell emphasized water's foundational role in addressing all humanitarian issues, making it the crisis requiring immediate global attention.

Chapter 6

Protecting Our Rivers

Keepers recently mounted yet another challenge to the government of Alberta, this time regarding changes to the Water Act, which will allow industry to transfer water between three river basins. The changes would provide easier access for oil and gas companies to northern Alberta's water.

Keepers contends that combining water from the Peace–Slave and Athabasca River basins would "threaten waterflow, spread invasive species, shift chemical balances, and degrade downstream environments." The three rivers being considered flow north, and according to Keepers, will affect the health and safety of Indigenous people.

"There is no justice in a process that excludes the people who live and rely on these waters," says Jesse Cardinal, executive director of KOW.

According to an article in the *Lakeside Leader,* Slave Lake's local newspaper, Alberta's Water Act prohibits transfers between basins without special legislation. The proposal to alter the river basins is being promoted by the Canadian Association of Petroleum Producers.[13]

Despite recent claims from Alberta and environment ministers about improved relations, Indigenous communities are continually excluded from having a seat at the table.

Jesse Cardinal, the passionate and persistent Executive Director of the KOW, says "There is no accountability in a system where decisions are made behind closed doors and framed as consultation."

Keepers do their best to inform their network of concerned citizens. Cardinal continues to educate the public so that Keepers' concerns become our concerns. She has called for severing the ties between the legislative government and oil and gas companies, and stresses that descendants of settlers have a responsibility to hold the government accountable.

The Keepers openly share details of their Community Water Monitoring Program. After years of discussing water monitoring data at their monthly meetings, it became clear that they would not likely ever obtain in-depth access to government and industry water monitoring data, feeling that it's doubtful it would be trustworthy even if it was shared.

The ideal solution would be for KOW to initiate a citizen-driven and citizen-based monitoring program in the Athabasca Watershed.[14] I identify with KOW's lack of trust of government. I feel this is well justified.

Keepers have created an interactive water monitoring map; the initial phase of this program collects essential water quality measurements. When the program grows, more types of measurements will be taken.

HOW ONE KID STOPPED
THE CONTAMINATION OF A RIVER

At eleven, Stella Bowles discovered Nova Scotia's LaHave River was contaminated by raw sewage from illegal straight pipes. Unable to swim safely, she tested water samples revealing dangerous fecal bacteria levels and publicized her findings through Facebook and warning signs.

Her persistent advocacy raised community awareness and prompted government action, winning her science awards and helping secure a $15 million commitment to replace all straight pipes with septic systems by 2023.

Her efforts achieved real environmental change, demonstrating how one determined young person can create significant impact. At sixteen, Bowles became the youngest recipient ever of the Order of Nova Scotia for her river cleanup work.

Chapter 7

Transport Canada

Late in 2024, a federal committee was convened to question cabinet ministers about Transport Canada's failure to notify Indigenous communities regarding water and soil contamination at a dock in Fort Chipewyan, Alberta. The federal Standing Committee on Environment and Sustainable Development approved a motion by New Democratic Party (NDP) MP Laurel Collins, calling for cabinet ministers, Indigenous leaders, and experts to testify about the contamination and the communication lapse.

"It is particularly egregious that the government knew about the contaminated dock at least since 2017 and that there haven't been steps to address it," Collins told Canada's *National Observer* in a phone interview. She emphasized that the health of First Nations must be prioritized.[15]

"Everybody seems to be baffled by it, and nobody wants to address it," Chief Alan Adam said. A few days after the three First Nations and Métis leaders went public about the contamination, Minister of Environment and Climate Change Steven Guilbeault told Chief Adam that the Minister of Transport Anita Anand, would get in touch with him. More than two weeks later, this had not happened, said Adam.[16]

In 2017, a Transport Canada study found that sediment and

groundwater samples near the dock, where community members swim and fish, contained nickel, arsenic, and harmful hydrocarbon residues at levels exceeding guidelines. Hydrocarbons are petroleum products, such as gasoline and kerosene, and are also ingredients in many paint thinners, solvents, glues, and some cleaning products. They are dangerous when the fumes are breathed or the liquid gets into the lungs. Both arsenic and nickel are known carcinogens and can have other adverse health effects. Indigenous leaders disclosed that Transport Canada had been aware of the contamination at the Fort Chipewyan dock and surrounding areas for years but chose not to inform the community.[17]

The Athabasca Chipewyan First Nation (ACFN) requested that the Transport Canada dredge the river around the dock, which serves as a crucial escape route during wildfire season. Transport Canada declined, so the ACFN hired a contractor to carry out the work. During the process, it was discovered that the agency had refused because dredging would disturb contaminants. Transport Canada has yet to share the health and ecological risk assessment with the affected communities or the media.

Collins emphasized that the federal government must be held accountable for its commitment to uphold treaty rights, implement reconciliation, and protect the health and well-being of Indigenous communities. She stated that failing to inform the Nations in this territory is an infringement on their aboriginal and treaty rights, disregards human health, and is a clear example of environmental racism.[18]

Collins' motion called for several cabinet ministers to explain the federal government's decision-making, including Minister of Transport Anita Anand, Minister of Environment and Climate Change Steven Guilbeault, Minister of Emergency Preparedness Harjit Sajjan, Minister of Indigenous Services Patty Hajdu, and Minister of Crown-Indigenous Relations Gary Anandasangaree. The ministers were questioned in a two-hour meeting with the committee. Included were ACFN Chief Allan Adam, Mikisew Cree First Nation Chief Billy-Joe Tuccaro, and Fort Chipewyan Métis Nation President

Kendrick Cardinal to explain how this incident is affecting their communities.

Chief Allan Adam said: "Canadians expect this industry to be properly regulated. I'm here to tell you that it's not true. It's not regulated."

Anand's press secretary at the time said the government "is working with the local First Nations community to ensure that operations at the port facility are carried out safely," in an emailed statement to Canada's *National Observer* on October 2, 2024. The community wants to deal with the issue, but the lack of action and communication "makes it frustrating for everybody," said Adam. "I don't want to say something bad in regard to this, but I just hope something good comes out of it."

I try to imagine walking a day in Chief Adams' footsteps. I wouldn't want to say anything bad or criticize any level of government either if I were him. I encourage you to visit my website, susanblacklin.com to watch a video of how police treated him in what I believe to be a horribly racist manner.[19]

Surely, I cannot be the only person sickened and ashamed by such behaviour.

Collins said it is appalling that Anand "hasn't reached out or contacted the nations who are impacted by this, especially after they gave an explicit call to action." The three leaders explained that not only is the Transport Canada dock critical infrastructure for day-to-day transportation and a lifeline during emergencies, it is also an area where community members swim and harvest food. Her motion additionally invites Mandy Olsgard, a toxicologist advising the nations, and Commissioner of the Environment and Sustainable Development Jerry DeMarco, to testify. Collins wanted research to begin as soon as possible, but the motion says the Fort Chipewyan dock study will begin after the environment committee finishes hearing witnesses for three other ongoing studies: factors that led to the recent fires in Jasper National Park, environment and climate impacts related to the Canadian financial system, and a briefing on the net zero accelerator fund.

"This is an urgent matter and an emergency for these communities," Collins said. "This is their only way of evacuating in the event of a wildfire, which we know has happened before in the area, and will happen again." Transport Canada spokesperson Sau Sau Lui told Canada's *National Observer* in an email statement, "the site was not likely to pose any risks to human health" based on the department's health and ecological risk assessment. Lui did not share a copy of the follow-up assessment or explain why the department would not make it public. Like many others, I would like to see the science that supports such a statement.

Transport Canada has reached out to representatives of the ACFN, Lui added. "It's going to take all levels of government—federal, provincial, and municipal—to address this issue."

Adam said that, for example, the community has been asking for the water intake to be moved about thirty kilometres up the lake ever since Adam was elected Chief in 2007. "People want to have that assurance, but it's going to cost quite a bit," he noted.

Chief Adams' community was faced with two challenges related to this Transport Canada grievance. First there is the contaminated dock area; I suspect the water intake is too close to this contaminated site. The second problem is if too many water license users take water from the Athabasca River, then levels drop significantly, restricting their treaty rights, but also their safety, for example, if they should need to flee wildfires.

Here is yet another example of how the Indian agent is dictating to the First Nations what will happen to the water on their own land.

CLIMATE IS PEOPLE AND THE ENVIRONMENT IS PEOPLE BECAUSE WE ARE SO CONNECTED TO OUR ENVIRONMENTS

Abhayjeet Singh Sachal, a University of Toronto student studying Global Health and Peace Conflict and Justice, founded Break the Divide to foster climate and mental health action.

After returning from Malawi, where he researched COVID-19 responses, he connected students from Surrey, BC, with peers in Inuvik, Northwest Territories. This exchange helped both communities understand each other's challenges—melting permafrost affecting Indigenous hunting versus urban issues like immigration and violence.

Break the Divide transforms apathy into action through facilitated conversations that build empathy across geographic barriers. Since 2017, the organization has expanded globally, connecting youth internationally while addressing multiple UN Sustainable Development Goals through personalized education on climate change and mental health.

Chapter 8

Clean-up Costs Add Insult to Injury

Alberta faces tens of billions of dollars in clean-up costs for oil and gas sites across the province. Perhaps hundreds of billions more if you count the oil sands mines in the north. It's been a growing problem for years. Now the Alberta Energy Regulator (AER) has released what it calls its first annual liability management industry performance report.

In 2022, there were 464,000 licensed wells across Alberta, and the number rose to 468,000 in 2023, according to the regulator. There are minimum amounts companies must spend each year cleaning up abandoned oil and gas wells. Companies were required by the regulator to spend $422 million in 2022, cleaning up orphaned oil and gas wells.

These may sound like eye-poppingly big numbers. In reality, they are a drop in the bucket compared to the latest estimate of total clean-up costs of $33.3 billion, not including the oil sands.

The regulator has also started tracking the financial health of companies in a new way, attempting to get a better sense of who can, and cannot, sufficiently manage liability obligations.

A revised government-commissioned report on Alberta's inactive oil wells now recommends that the province oversee a new insurance fund, instead of relying on taxpayer money to guarantee it.

This change in wording is among several updates made after a draft was leaked, which sparked concerns that the government might allow oil and gas companies to avoid their legal responsibility for well reclamation. The report, written by David Yager (a former oilfield services executive and current AER board member) offers more than twenty recommendations to address the nearly 80,000 inactive, orphan oil and gas wells scattered throughout Alberta.[20]

The Alberta government has returned more than $137 million to Ottawa after failing to use the funds in time for cleaning up abandoned oil and gas wells. Despite the urgent need with tens of thousands of inactive wells remaining, provincial authorities couldn't deploy the much-needed support, leaving many companies frustrated by the missed opportunity for vital reclamation work.

This money was part of the federal government's $1.7 billion initiative launched in 2020 to tackle environmental risks from aging energy infrastructure and to provide jobs for oilfield workers hit hard by the pandemic and plummeting oil prices. The funding was split between British Columbia ($120 million), Alberta ($1 billion), and Saskatchewan ($400 million), with Alberta's Orphan Well Association also receiving a $200 million loan to address wells abandoned by bankrupt companies.[21]

Further muddying the statistical waters was the federal funding for what was dubbed the Site Rehabilitation Program, which incentivized more clean-up activity in 2022 and makes it difficult to strictly assess the effectiveness of the regulator's new framework. That program provided a total of $1 billion over three years to oilfield service companies conducting clean work.

A municipality in Northern British Columbia is grappling with a severe financial crisis after an oil company collapsed owing nearly $9 million in unpaid property taxes, interest and penalties. Erikson National Energy became insolvent in 2024 and entered receivership in May 2025, accounting for almost all of these outstanding debts. With no realistic chance of recovering the money, the Northern Rockies Regional Municipality (NRRM) has formally requested provincial approval from the Minister of Housing and Municipal Affairs to write off $8.7 million, warning that additional write-off requests may follow.

"This certainly is one of the largest write-off requests that we've made," Toni Pike, NRRM's director of finance, told council members in October 2025, cautioning that more losses could be coming.

The situation is complicated by Erikson's operations being on Crown land leases, which cannot be sold to recover funds, according to the provincial government. Council was also warned that other bankrupt companies holding Crown leases have similarly failed to pay taxes or may soon default.

NRRM Mayor Rob Fraser expressed concern about the municipality's future, given its heavy reliance on industrial tax revenue. Questions remain about whether shareholders received dividends before the company's collapse and why no contingency fund existed to protect against such losses.[22]

Canadians should not be comfortable footing the bill for toxic clean-up when those with investments in these companies have reaped massive rewards. But we can't blame the investors; they took this money with the full consent of the Canadian government.

Nelson, British Columbia, is joining ten other British Columbian municipalities in a proposed class action lawsuit that aims to hold oil companies responsible for climate disaster costs. Slocan, Burnaby, Cumberland, Squamish, View Royal, Sechelt, Gibsons, Qualicum Beach, Port Moody, and Pemberton have already signed the suit, declaring that communities can no longer manage the skyrocketing costs of climate change. West Kootenay, Climate Hub, and West Coast Environmental Law are calling on other local governments to commit to this groundbreaking lawsuit.[23]

How does climate change impact water quality? Climate change is attacking water quality from every direction. Warming waters suffocate aquatic life and spawn toxic algal blooms. Extreme rainfall floods our waterways with pollutants while droughts concentrate contaminants to dangerous levels. Rising seas are salting our freshwater supplies. Hurricanes overwhelm sewage systems, dumping untreated waste into drinking water. Melting glaciers threaten billions who depend on them. Wildfires leave landscapes that poison reservoirs for years. These threats aren't isolated—they're compounding. The water crisis is here.

I once worked at a prestigious advertising agency in the UK called Hedley Byrne, where many offices had corner windows looking out to Park Lane and Oxford Street. Its largest account was Caterpillar equipment, and they created advertising campaigns for the oil fields in Canada. I had no idea at the time, how far reaching the profits of big oil travelled. Neither was I aware of the oilfields significant negative impact on the land, water and lives of so many Indigenous people.

The Watershed Sentinel published an article in their October/November 2024 issue, titled "Alberta's Emergency: Auditor blows the whistle on water mismanagement."[24] Unless the province overhauls its systems, the Auditor General of Alberta's July 2024 report on Surface Water Management predicts parts of the province may lose access to fresh, clean drinking water.

The article reveals that the Department of Environment and Protected Areas has failed to set water conservation goals in most watersheds and therefore cannot gauge whether existing water conservation plans are working. It indicates the agency lacks the capacity for water pressure monitoring, risk assessments, and setting water conservation objectives. It details how water license approval and monitoring processes are ineffective and not consistently enforced.

The article stated that the industry's own data related to pollution from mining operations is ineffective, and are not consistently enforced. Money talks, and those with big money are still making the rules. This must stop.

WINNIPEG TEEN TO REPRESENT CANADA AT YOUTH STEM COMPETITION

Winnipeg teen Keerthan Kamala Krishnamoorthy was selected as one of eight students representing Team Canada at the 2023 Regeneron International Science and Engineering Fair (ISEF).

The sixteen-year-old Fort Richmond Collegiate student developed a mobile app using deep learning to detect harmful microorganisms in drinking water. His project utilized an inexpensive paper microscope to magnify water samples, allowing phone cameras to analyze safety.

A panel selected participants based on ingenuity, critical thinking, and scientific excellence. Keerthan planned to collaborate with University of Manitoba researchers and Lake Winnipeg Research Consortium to refine his innovation. He aspired to study computer science and machine learning while creating solutions benefiting people globally.

Chapter 9

Water Quantity

Alberta has 25,000 water license holders controlling 9.5 billion cubic metres of water, yet even in severe droughts, the province cannot compel them to share or conserve. With all of southern Alberta's water already allocated and no law to enforce sharing, the government began unprecedented negotiations in early 2024 with major license holders, aiming for significant, rapid reductions in water use. This marks the largest water-sharing effort in Alberta's history, as officials urge cooperation to address the urgent crisis.[25]

I would like to share an example of what this means. If an oil and gas company with an old licence (say from 1950) has rights to take fifty million litres per day, and a town downstream that got its license in 1980 needs water for 10,000 residents, then during drought, the government must let the company take its full amount before the town gets any water. Even if people's taps run dry, the company cannot be forced to reduce usage or share; the long-standing water licences rule!

Tailings are the waste materials left over after the extraction of valuable minerals from ore during the mining process. These materials, which often consist of finely ground rock and chemical by-products, can pose significant environmental hazards if not managed properly. ACFN is taking legal action to hold the AER accountable for toxic tailings spills that have contaminated key areas of its traditional territory, including the Firebag River Corridor.

This is a clear violation of the Treaty 8 agreement, and a disruption to its ability to assert sovereignty and practice Dene law on its land. The toxic tailings spill violates Treaty 8 because it contaminates the Firebag River and traditional territory, directly infringing on the ACFN's constitutionally protected rights to hunt, fish, and gather in a healthy environment.[26] The AER is also accused of breaching the treaty's spirit and substance by failing to inform the community, consult on mitigation efforts, and effectively regulate the spill, thereby harming the ACFN's traditional way of life.

Treaty 8, signed in 1899, is the largest of Canada's numbered treaties by area, covering parts of northern Alberta, northern Saskatchewan, northeastern British Columbia, and southern Northwest Territories. It was negotiated to extinguish Indigenous title to the land in exchange for peace, friendship, and certain rights, including the continued ability to hunt, trap, and fish, subject to regulations, and the provision of resources like agricultural implements to support a transition to farming.

ACFN's challenge is the aftermath of discharged toxic chemicals from the Kearl Oil Sands Processing Plant and Mine (the Kearl Facility.) There were three uncontrolled discharges of industrial wastewater at the Kearl Facility from May 2022 to November 2023, resulting in over 5.3 million litres of tailings and 670,000 litres of other contaminated water being released in the territory. For nine months, the AER knew about ongoing uncontrolled discharges of tailings at the Kearl Facility and chose not to inform ACFN. During that time, the Dene community continued to harvest food, hunt, fish, and trap, and use water from the polluted watershed.

This is not just a spill—it is a symptom of something shameful and deep-rooted in Canada. The choice not to inform ACFN of the spills is another example of a broken system that ignores the value of Indigenous lives and does nothing to encourage reconciliation.

CORB LUND PETITIONS AGAINST COAL

The Juno Award winner, who ranches near the Old Man River downstream from proposed mining sites, has opposed coal mining for over five years. His petition seeks to legally prohibit all new coal mining activities, approvals, and permits in the region.

The Alberta Government changed the rules, requiring he submit his citizen initiative again which he did. The government changed the rules again after the legislature was closed (December 2025) and announced the fee to initiate a petition was being raised from $500 to $25,000. However, they missed the part that said if a petition had already been paid and was being rewritten, no new fee would be necessary.

Luckily, the No More Coal petition does not need to raise that kind of money. I don't doubt for a minute that further tricks will be attempted. I don't know how we stop this slide into autocracy unless enough people become advocates.

Lund was inspired to launch the petition after Premier Smith suggested it at a rowdy town hall in Fort Macleod, where residents opposed coal mines. He has four months to collect 177,000 signatures—though he aims for double that amount to demonstrate overwhelming public opposition.

For Alberta residents only: https://www.coalpetition.ca

Chapter 10

Coal in the Water?

AER's recent approval of coal exploration on the eastern slopes of the Rocky Mountains is drawing quick reaction from landowners, mayors, and community groups across southern Alberta.[27] The debate over the revival of coal development in the region has been contentious, with supporters arguing it could bring economic development to communities that need it. Those opposed to the project raise concerns around long-term risks to land and water.

The AER approved Northback Holdings (with conditions) to conduct a coal exploration program at the Grassy Mountain site, located in the Municipal District of Ranchland. In a written decision, the regulator said the applications met all regulatory requirements and that the exploration program was in the public interest. It added that the applications weren't for a coal mine, but for an exploration program at the existing, un-reclaimed Grassy Mountain mine site.

"The exploration program is designed to improve Northback's understanding of the extent of the Grassy Mountain coal deposit, obtain raw coal samples, and structurally model the coal seam complexities," the decision reads. It added that if Northback proceeds with mine applications at Grassy Mountain in the future, it must follow a "rigorous regulatory process that all resource development applications must follow," and that accepting the exploration applications "does not constitute approval of a coal mine."

What is this "rigorous regulatory process?" How can anyone claim that the Alberta regulations are stringent? Alberta's environmental laws and regulations, including their drinking water guidelines, have been the weakest of all provinces for decades. The weak, often biased system that was in place has now been wiped away. Replaced by Bill 15, which has First Nations, local governments, legal experts, environmental advocates, and opposition Members of the Legislative Assembly (MLAs) warning of troubling government overreach. Later in the chapter titled "Bill Who," I will explain the similarities and pitfalls of Bill 5 in Ontario, Bill 15 in British Columbia, Bill 54 in Alberta, and the federal Bill C5.

Northback Holding exploration at the Grassy Mountain site was previously rejected in 2021 by the federal government, which said the project would likely cause significant adverse environmental effects, including on surface water quality, including from selenium effluent discharge." In its opposition, the district of Ranchland had noted environmental concerns tied to selenium but then stated in its recent decision that those concerns didn't apply to the exploration work.

What didn't apply? Any water standards regarding selenium? That the regulator doesn't believe selenium is an issue in drinking water? Are those in charge suggesting that selenium is not a problem during exploration, only in production? Mining operations pose environmental risks related to selenium throughout exploration and production stages, with the issue typically becoming more severe once active extraction begins. While selenium exists naturally in the environment, industrial activities speed up the process by which selenium-containing geological materials enter aquatic ecosystems.

Still, some landowners and environmental advocates say the development sets the stage for coal mining in a critical watershed that provides drinking water to more than 200,000 people.

"The elephant in the room is that this is not about an exploration permit. It's about the next step or ticking the box to do the next steps toward going for a mine," said Norma Dougal, who is on the board of directors for the Livingstone Landowners Group.[28]

"What we are totally expecting is to have to go back and waste our time at a full-blown hearing, when it's already been shown that a mine at Grassy is not in Alberta's best interest, economically or environmentally."

Davis, the Reeve of Ranchland, said he expected that Northback wouldn't be conducting exploration if they didn't have interest in opening a coal mine. He added the municipal district continues to hold environmental concerns.

"Southern Alberta is a very water-poor area. And the loss of any water sources or watersheds are detrimental to southern Alberta, particularly," he said. "And of course, that goes along with the other problems of contamination of the water resources that we have, which are meagre at best."

Settler democracy will have to step back in many areas as Indigenous communities re-assert their jurisdiction.[29] That's one reason we should all support the British Columbia Declaration on the Rights of Indigenous Peoples Act, and the important work of updating our laws to comply with United Nations Declaration on the Rights of Indigenous Peoples (UNDRIP). When it comes to mega dams and pipelines, power is not a metaphor. The people who control our energy control us—and they know it.

ONTARIO BECAME THE FIRST NORTH AMERICAN JURISDICTION TO COMPLETELY ELIMINATE COAL-FIRED ELECTRICITY

At the turn of the century, David Devereaux monitored Ontario's coal-fired power plants from the control room of the province's Independent Electricity System Operator. The Lakeview Generating Station in Mississauga and four other coal plants across Ontario belched toxic smoke—mercury and sulphur dioxide—that darkened southern Ontario's skies with deadly smog, especially during hot summers when air conditioning demand surged.

These plants supplied a third of Ontario's electricity but posed a severe environmental and public health crisis. By 2001, pressure mounted to shut them down, though officials warned there was "no easy off button."

Yet, in 2014, Ontario achieved what seemed impossible: it became the first North American jurisdiction to completely eliminate coal-fired electricity. This landmark decision removed 17 percent of the province's emissions —equivalent to taking seven million cars off the road— and became the continent's largest greenhouse gas reduction initiative to date.

Chapter 11

$3.6 Million Fine: A Drop in the Bucket

Elk Valley Resources Operations Limited (EVR), formerly Teck Coal Limited, has been slapped with a $3.6 million fine for contaminating southeastern British Columbia's waters with toxic selenium and nitrate.[30] While that might sound substantial, consider this: the company made over $13 billion in profit during the same period these violations occurred. As Simon Wiebe of the conservation group Wildsight bluntly put it, the penalties are nowhere near large enough to deter further pollution.[31]

The seven penalties issued by Environmental Management Act director Jason Bourgeois document violations from 2018 to 2023 across five coal mining operations. The company's most egregious failure? Refusing to build three mandated water treatment facilities by their deadlines, despite a 2014 ministerial order requiring action on contamination that has plagued the region since the mid-1990s.

Between March 2021 and March 2023 alone, EVR violated provincial limits an astounding 173 times—forty-seven selenium breaches and 126 nitrate violations. This comes on top of a record $16 million penalty in 2023 for earlier pollution failures. When Glencore acquired Teck's coal operations in 2024, they inherited not just profitable mines, but a legacy of environmental destruction that EVR is now appealing.

Kathryn Teneese, chair of the Ktunaxa Nation Council, called the ongoing non-compliance "unacceptable." For the Ktunaxa people, the polluted Elk and Kootenay rivers aren't just waterways, they're essential for harvesting rights and the continuation of cultural and spiritual practices spanning generations.

The contamination has crossed international borders, with Canada repeatedly violating the 1909 Boundary Waters Treaty as toxic waters flow into Montana.[32] Tom McDonald of Montana's Coast Salish Kootenai Tribes described fish so deformed they're missing gill plates and unable to spawn. "We don't feel comfortable eating the fish there anymore," he said. "That shakes you up quite a bit."

A 2023 US geological survey study documented selenium increases in the Elk River as the largest ever recorded worldwide in peer-reviewed research—a shameful distinction for Canadian environmental management.

Here's where the story gets truly alarming. While researching selenium's health effects, I spotted this disturbing notice appears on the National Institute of Health's website: "Because of a lapse in government funding, the information on this website may not be up to date." The irony is staggering; governments find money to subsidize pipelines, fracking, and coal mining, but can't fund the basic sharing of critical information like health monitoring for citizens.

Selenium exposure causes selenosis in humans, characterized by hair loss, nail deformities, garlic-scented breath, cognitive decline, and gastrointestinal disorders. Research links chronic high-level selenium consumption to prostate cancer and neurotoxicity. In severe cases, exposure can be fatal.[33]

In 2013, published, peer reviewed work by Marco Vinceti, Catherine M Crespi, Francesca Bonvicini, Carlotta Malagoli, Margherita Ferrante, Sandra Marmiroli, and Saverio Stranges recommend drinking water limits for selenium of 1 µg/L to reduce risks of cancer, endocrine disorders, and neurological diseases. Canada's maximum acceptable concentration (MAC)? A shocking 0.05

mg/L that is 50 µg/L—fifty times higher than what unbiased scientific research recommends as safe.[34]

I don't believe this is just a regulatory failure; I believe it's a deliberate choice to prioritize industry profits over public health.

EVR boasts of "significant progress," claiming its four treatment facilities remove 95 to 99 percent of selenium from treated water after investing $1.5 billion. They project another $700 million in spending by 2027. Yet according to director Bourgeois's decision, full compliance won't happen until 2028.

If it continues on track, then that is another four years of communities and ecosystems suffering preventable contamination. Meanwhile, Canada and the United States have requested International Joint Commission intervention, with recommendations expected in September 2026, which carry no binding authority.

The message is clear: inadequate penalties, weak drinking water standards, and non-binding international oversight create a system where pollution pays and communities pay the price.

Worldwide solar and wind power generation has outpaced electricity demand this year, and for the first time on record, renewable energies combined generated more power than coal, according to a new analysis.[35]

Global solar generation grew by a record 31 percent in the first half of the year, while wind generation grew by 7.7 percent, according to the report by the global energy think-tank Ember.

Solar and wind generation combined grew by more than 400 terawatt hours, which was more than overall global demand increased in the same period.

Renewables accounted for 34.3 percent of electricity generation globally in the first half of 2025, while coal accounted for 33.1 percent. It's the first-time renewables have overtaken coal, according to the report.

The findings suggest it is possible for the world to wean off polluting sources of power, even as demand for electricity skyrockets, with continued investment in renewables including solar, wind, hydropower, bioenergy and geothermal energies.

What I find most discouraging is the about-turn that elected officials make as soon as the election is over. A decade ago, Mark Carney made a rousing speech about how politicians' shortsightedness gets in the way of urgent climate action; he called it the "tragedy on the horizon." And yet, just months into his political career, he's killed the carbon tax, paused a mandate for electric vehicles and championed the liquefied natural gas (LNG) industry.[36]

I have no doubt in my mind that Mark Carney is the best man to lead Canada in these difficult times with our southern neighbour. Who could have foreseen that he would become the pipeline guru? What about all his prior involvement, and that of his better half, Diana Fox Carney, in sustainable energy, environmental protection, and climate change? I can only hope he is buying time and will soon place significant value on climate change, renewable energy, and above all, protecting our source waters and providing all Canadians with safe drinking water.

Each time I take a ferry to or from Vancouver, I see coal dust that coats nearby Tsawwassen condos with dirty black grime, preventing owners from opening their windows or sitting on their patios. Streams of toxic waste pour into Burrard Inlet, home to Canada's largest port. These waters, at the northern extent of Vancouver, have nourished the Tsleil-Waututh Nation (səlilwətał) for millennia.

Today, they are so polluted that public officials say shellfish are no longer safe to eat, and people are regularly barred from swimming at local beaches.

SEVENTEEN-YEAR-OLD HALTS PLANS FOR SOLID-WASTE INCINERATOR TO BE BUILT IN HER NEIGHBOURHOOD

Growing up in Baltimore's Curtis Bay neighbourhood, Destiny Watford witnessed many friends and family members suffering from asthma. She connected the high asthma rates to air pollution from a local medical-waste incinerator, coal pier, and medical plants.

At seventeen, when a new solid-waste incinerator was scheduled near her high school, Destiny and other young people pressured officials to stop it. After three years of activism, state regulators revoked the project's permit in March 2016.

Her efforts earned her the Goldman Environmental Prize for highlighting environmental inequities. As a college student, Destiny worked for Free Your Voice, advocating to reinstate air quality monitors in Curtis Bay.

Chapter 12

Other Contaminants

It is not just big oil that threatens our source waters, and it is not only Alberta where we have a problem. There's the cocktail of contaminants that rain washes in from city streets and the sewage that pours in during big storms. There is the mix of heavy metals and toxic chemicals the government allows corporations to dump into the water every day: oil and grease, lead, copper and more, melding with cloudy wastewater threatening marine life.

Today we are hearing of atmospheric rivers and heavy rains that were never mentioned six or seven years ago. A continent-wide investigation has revealed thousands of landfills across the UK and Europe are located in floodplains, risking toxic waste release into water supplies and ecosystems as climate change intensifies flooding.

While Canada professes to enforce strict pollution control regulations for landfill sites, which often require protective lining systems, I have not been able to find one example of such enforcement. As Canada is typically twenty or more years behind the UK and Europe, I have little faith this isn't happening here in Canada too.

Researchers warn that roughly 90 percent of Europe's estimated 500,000 landfills, including 22,000 UK sites, predate pollution control regulations like protective linings. Climate-driven floods could release plastics, toxic metals, PFAS "forever chemicals," and PCBs into groundwater and food chains. More than half the mapped sites are in areas where groundwater already fails chemical quality standards.

The investigation, conducted by *The Guardian*, Watershed Investigations, and Investigate Europe, exposed a critical gap: the EU lacks centralized landfill records, and member state data is fragmented and inconsistent. Experts emphasize the uncertainty is alarming; while most landfills may be safe, even a small number containing highly toxic materials could pose serious threats, and authorities don't know which ones.[37]

Between 1971 and 2016, Burrard Inlet, British Columbia, was contaminated by 700 different pollutants, as reported by the Tsleil-Waututh Nation. Hydrocarbons coat the seafloor, pesticides poison fish and shellfish, and heavy metals saturate the water. At least twenty-four contaminants reach levels that make harvesting food from the inlet hazardous. This devastation marks a stark violation of a once-thriving ecosystem that sustained generations. Tsleil-Waututh Nation is determined to challenge the rules that allowed such destruction, fighting to restore what was lost and protect their ancestral waters from further harm.[38] This is just the pollution we know about.

This raises many questions: How often do spills and leaks occur along Burrard Inlet? How many go unreported? And, most importantly, what will it take to demand change? This is not a provincial problem, similar problems occur in Alberta and other provinces, reaffirming why we need national drinking water regulations that are based on source water protection.

The AER granted approval to a disputed coal exploration initiative on the eastern Rocky Mountain slopes in May 2025. The Grassy Mountain project, operated by Northback Holdings Corp., had been denied in 2021 after a panel concluded that anticipated environmental harm to aquatic life and water resources would exceed any economic advantages.

Situated at a former coal mining site within the Municipal District of Ranchland, the project was resurrected two years following its initial rejection. In the previous year, it received an exemption from Alberta's prohibition on open-pit coal mining operations, as Northback's submission was deemed an "advanced" application. According to the regulator's written ruling, officials concluded that project approval serves the public good and that operations will not adversely impact water quality or animal populations, contrary to concerns raised by numerous participants during public consultation sessions.

Crowsnest Pass Mayor Blair Painter welcomed the regulator's decision and referenced Northback's promise that the mine would employ roughly 300 people should it move ahead.[39]

"We all want clean water. I want clean water. All the residents of Crowsnest Pass want clean water," he said in an interview.

I can only assume he wants jobs more than he wants safe drinking water. He continued, "If Northback is successful with moving ahead with the mining application, then we can feel confident that they will do this within the regulations of the province and the federal government when it comes to clean water."

I wonder what regulations for water, either federal or provincial, he is referring to. If such regulations existed, then surely the Keepers would not have spent the past three decades (or longer) demanding safe drinking water for all the communities downstream of the oilsands and Swan Hills toxic waste site. This is a problem when politicians refer to laws or protection that simply do not exist or, if they do exist, are not enforced. It gives the general public a false sense of security that they are in good hands.

PRICE OF WATER

British Columbia's government is failing to hold industries accountable for exploiting public water resources. Oil and gas companies, smelters, and bottlers extract massive quantities while paying virtually nothing—shortchanging the province tens of millions annually.

If BC matched Nova Scotia's fees, it could collect an additional $142 million yearly for desperately needed watershed restoration. The BC Watershed Security Coalition demands the government raise water rental fees and invest proceeds in protecting damaged ecosystems. "Aligning BC's rates with other provinces would generate funding to protect our water—without burdening households or small businesses," they argue.

Citizens must pressure officials to stop giving away this vital public resource.

Chapter 13

Trihalomethanes (THMs)

I am not a scientist, but I like to offer this example of guidelines versus regulations. This false security that the government repeatedly offers to Canadians can be life changing.

Trihalomethanes (THMs) are a dangerous group of chemical compounds formed when chlorine, used to disinfect drinking water, reacts with natural organic matter in the water. You don't have to drink THMs. Inhaling them can be fatal. This leads me to a recent statement submitted by the Safe Drinking Water Foundation (SDWF), which is a charity, to a government organization that, in my opinion, should know more than the charity.

I heard Dr. H_2O repeatedly say that the main concern regarding THMs is from inhalation. And the problem is most likely created by WTOs who dump another jug of chlorine, maybe it's the "if one jug is recommended then two jugs should be even better" mentality. Or maybe it's to try and meet another, more frequent test, such as Coliforms?

THMs are formed when organics react with chlorine. Chlorine is used to kill all the parasites/bacteria/protozoa/etc. and it reacts with the organic material (the crud in the water) and forms THMs. There are ways to remove THMs. Those who can't afford the point-of-use treatment systems (the people who are most likely to become ill) are also those who will have more trouble affording the medications if they don't treat their water.

This is the financial argument for water regulations. The government approach of "keep them as low as reasonably achievable" and "people can use treatment systems to remove some of it" does not protect human health. THMs are not regulated and not everyone has the knowledge and the financial ability to remove them.

The statement below is from an archived website of the National Collaboration Centre for Environmental Health. Finally, a government agency speaking the truth in plain English:

Information about Canadian drinking water systems and past waterborne disease outbreaks is incomplete and non-standardized. Standard definitions and coordinated surveillance systems for waterborne disease outbreaks would help inform policy and practice.

A relatively high proportion of past waterborne disease outbreaks in Canada are estimated to have occurred in small drinking water systems serving populations of 5,000 people or less. Waterborne disease outbreaks in small drinking water systems are often the result of a combination of water system failures; contributing factors often include a lack of source water protection and inadequate drinking water treatment.[40]

Analyses suggest small drinking water systems face challenges associated with infrastructure, technology, and financial constraints. Investments in drinking water systems and operator training have the potential to reduce the burden of waterborne disease in Canada.

Nicole Hancock, executive director of the SDWF, does an excellent job of advocating for this. Here is part of the recommendations which Hancock submitted to representatives of the Water and Air Quality Bureau of Health Canada and the members of the Federal-Provincial-Territorial Committee on Drinking Water: "We are very concerned about the fact that a guideline value for the maximum acceptable concentration (MAC) for trihalomethanes (THMs) is being proposed and not a Canadian regulation or standard. Some THMs have been classified as probable human carcinogens by the International Agency for Research on Cancer. Several studies have linked THMs to bladder cancer risk, and the Occupational and Environmental Branch of the Division of Cancer Epidemiology & Genetics at the National Cancer Institute has described a new association between disinfection by-products and endometrial cancer."

While Canada's proposed MAC for THMs is the same as the standard that the United States Environmental Protection Agency (EPA) has set (one hundred parts per billion), the difference is that the proposed MAC for THMs in Canada would be a guideline, whereas, in the United States, it would be a standard, which means that it is legally enforceable. In fact, Canada is one of only two developed countries that do not have legally enforceable regulations or standards for drinking water (the other country being Australia). Australia's guidelines were called an international embarrassment by Australian chemical risk expert Mariann Lloyd-Smith who stated, "Australia cannot continue to use drinking water guidelines that are an international embarrassment."[41]

On the Government of Canada's Draft Guidelines for Canadian Drinking Water Quality Trihalomethanes webpage, it is stated that "utilities should make every effort to maintain concentrations as low as reasonably achievable without compromising the effectiveness of disinfection." This is somewhat similar to the Guideline Technical Document for Arsenic, in which it is stated that the MAC "...for arsenic in drinking water is 0.010 mg/L (10 µg/L) based on municipal, residential scale treatment achievability. Certified residential treatment devices are commercially available to remove arsenic to well below this concentration. Every effort should be made to maintain arsenic levels in drinking water as low as reasonably achievable (or ALARA)."

However, it only took me one minute on the website https://waterquality.saskatchewan.ca/ to find a community in Saskatchewan that has a current water test result for arsenic that is nearly double the guideline (Buena Vista, tested January 21, 2025, at 16.7 µg/L). Despite the water being over the guideline for arsenic, most likely nothing will be done, and the level of arsenic won't even be tested for two years—what if it increases from that level? Also, I am certain that I could find communities in Saskatchewan with current water test results for arsenic that are even higher than that. Arsenic is also carcinogenic.

However, residents could purchase devices that are commercially available to remove arsenic to well below the guideline—if they had the financial capacity to do so and they were aware of the issue, knew the level of arsenic in their community's drinking water, and knew which commercially available devices lower the level of arsenic.

My first year of teaching, I taught in Morse, Saskatchewan. I do not have a background in science (I am a high school math and French teacher), and I did not know much about water at the time. The water coming out of the tap in Morse, Saskatchewan, was yellowish-brown, it smelled bad, and it tasted bad. I did not know that poor quality drinking water could cause cancer (yes, I have learned much during my over eighteen years with the Safe Drinking Water Foundation), so I only thought about how aesthetically unappealing the water was.

I used a commercial pitcher-style filtration system and filtered the water twice before drinking it. This was over twenty years ago, so I doubt that filtration technology was able to remove some of the parameters that it is now available to remove.

I fear every day for young teachers who are starting their careers in small towns across the country, do not know much about water, and might, every day, be ingesting tap water that could cause them to develop cancer or a permanent, neurological condition such as manganism.

The same Canadians who have difficulty affording the commercially available devices that remove these contaminants from their water will be the same Canadians who will find diseases that limit their ability to work and cause them to need medications extremely difficult. Furthermore, the already overburdened healthcare system will have another patient with cancer, leading us to wonder whether it is cheaper to treat the water properly or treat the cancer patient—we know the choice that the cancer patient would prefer!

Canadians not only drink tap water, but they shower and bathe with it as well. Individuals can be exposed during showering to elevated concentrations of chloroform from chlorinated tap water, through both dermal and inhalation exposure. Exposure to THMs in shower water may increase the risk of certain cancers, including bladder, colon, and rectal cancer. One study found that long-term exposure to THMs was associated with a twofold risk of bladder cancer, while another study of exposure to THMs during showering in Ottawa, Hamilton, and Toronto predicted thirty-six cancer incidents from exposure to THMs during showering.

These are major cities in Ontario, and the number is fairly small, but since they are major cities they also most likely have the benefit of their tap water being of high quality and the THM level being low. In fact, Toronto Water treats, stores, and distributes over 435 billion litres of safe potable (drinkable) water annually to all industrial, commercial, and household water users in the City of Toronto. Their accredited lab tests the Toronto drinking water every six hours (over 6,000 times per year), conducts more than 20,000 tests at the water treatment plants annually, and conducts 15,000 bacteriological tests on samples collected from the water distribution system annually. Despite Toronto Water's accredited lab testing the tap water for total THMs seventy-five times (and eleven times at the end of the line) during 2023 and having a maximum result of 16.3 µg/L of total THMs, Toronto contributed the highest number of possible cancer incidents (22) from exposure to THMs during showering.

This is approximately one out of every 137,545 Torontonians. What if, like with arsenic, a small town had double the guideline value in terms of concentration of THMs? Toronto has less than 17 percent of the guideline value in terms of concentration of THMs. The exposure to THMs during showering causes one cancer incident per 137,545 people. Who would want to take that risk? Some people who have the financial means and knowledge of the issues might use carbon filters that can remove THMs from drinking water.

However, that would not remove THMs from the water in which they shower. Also, some people who are very knowledgeable and concerned about their tap water quality might install whole-home reverse osmosis systems that can remove THMs. However, I do not foresee a teacher who moves to a small rural community for their first teaching job installing a whole-home reverse osmosis system. Canada is having great difficulties attracting health professionals to rural communities. Health professionals are knowledgeable about health—is it any wonder they do not want to live somewhere that would cause increased risks to their health?

MACs for parameters that have health considerations, including THMs, that are national, legally enforceable regulations or standards, at the lowest level possible, will safeguard the health of the greatest number of Canadians. If Canada does not wish to continue to be an international embarrassment, these national regulations or standards must be implemented as soon as possible."

That a charity (SDWF) needs to point this out to a regulating government organization is testament to my earlier point: charities exist because the government fails to do what we elected them to do, which is to protect all Canadians.

I hope I don't need to clarify this, but just in case anyone is wondering, the SDWF has never received funding for specific projects or to deliver projects from any government sources, other than small amounts through summer jobs programs, other employment programs, and benefits that applied to nearly all organizations during the COVID-19 pandemic—which I believe allows them to speak freely and honestly in the best interests of all Canadians.

I also wish to clarify that neither Dr. H_2O nor I have ever received as much as $1 from the charity. I am only aware of one other charity that has applied scientific expertise to consistently advocate for improved drinking water quality on behalf of all Canadians: the David Suzuki Foundation.

CALL FOR CIVIL DISOBEDIENCE

Activists in Alberta's Kananaskis Country are defending their fundamental right to clean water by opposing logging plans that threaten the Highwood River and endangered trout habitat.

These eastern slopes are critical headwaters for southern Alberta and the Prairies yet they face multiple threats from industrial activity. "We need more citizens prepared to stand up," says organizer Michael Sawyer, whose own tap water flows from these hills.

The group is fostering a culture of civil disobedience, recognizing that protecting watersheds requires direct action. They're calling on people to get involved, keeping watch and ready to physically defend the land that sustains entire communities downstream.

Chapter 14

One More Resource to Extract

Lithium mines are mainly found in Chile, Argentina, and Bolivia, (collectively known as the Lithium Triangle), while Australia, Chile, and China are the top producers. The Lithium Triangle countries have the largest reserves. I never considered that lithium would be mined in Canada.[42]

However, the demand for lithium is driven by batteries for electronics and electric vehicles, followed by ceramics and glass.

Some communities have more than their share of issues to fight. First Nations communities in Alberta are understandably concerned, as I estimate the majority still lack safe drinking water due to the Swan Hills Toxic Waste site and oil and gas companies. Yet, companies and research groups in Alberta are testing direct lithium extraction technology.[43]

This technology allows access to Alberta's lithium-brine potential found in existing oil and gas reservoirs, enabling the recovery of lithium concentrate, which is the first step in the value chain. Specifically, Lithium Bank Resources Corp. announced the acquisition of a wellbore (an existing drilled well that provides access to subsurface brine) within its 100-percent-owned Boardwalk Lithium Brine Project located in west-central Alberta, Canada. It is early days, and I wonder about due process, the right for First Nations to be consulted, and how this will impact their drinking water.

Lithium plants use vast amounts of water, often deepen existing shortages in the region. Argentina's poorest regions, comparable to Canadas Indigenous communities, has spurred demonstrations and lawsuits. In Argentina, too, they lack a formal process for negotiations between Indigenous communities and mining companies. Canadians are often reluctant to negotiate trade deals with China, citing human rights violations. Demand for your phones and tablets drives this consumer tech. In affected Chinese villages, it's everywhere. Villagers living near graphite companies are subjected to sparkling night air, damaged crops, homes and belongings covered in soot, polluted drinking water, along with government officials inclined to look the other way to benefit a major employer. After leaving these Chinese mines and refineries, much of the graphite is sold to Samsung SDI, LG Chem and Panasonic, the three largest manufacturers of lithium-ion batteries.

I doubt people the world over are prepared to give up their phones and tablets. There is insufficient Lithium available yet for recycling, so we still rely on it being mined, either from playa anhydrites (sedimentary deposits of anhydrous calcium sulfate) in environmentally and ecologically sensitive areas, or from deep mines with very toxic processes in countries where human health concerns are of little value. China, Chile or Canada: when are we going to address our own human rights issues here in Canada? Indigenous people deserve better, much better!

Elections are not the only time to think about our democracy. Each day is an opportunity to hold politicians accountable and to organize communities to stand up against short-sighted decisions. How do we achieve that? That is the big question.

WHO IS RESPONSIBLE?

Greater Victoria's water commission proposes making developers cover new water infrastructure costs as a "growth-pays-for-growth" policy to avoid burdening existing residents with rate increases.

Critics warn this adds thousands per home during a construction slowdown already strained by high costs and interest rates. "Housing projects basically do not financially work," says planning professor Mark Holland.

Meanwhile, water rates face double-digit increases starting in 2028 to fund nearly $2 billion in climate-related infrastructure.

The debate centres on accountability: Should current residents subsidize growth, or should developers (and ultimately new-home buyers) pay? With water security requiring massive investment, the question is who bears responsibility for sustainable infrastructure.

Chapter 15

Water Justice

In Winnipeg on my book tour in spring 2024, I was thrilled to see a poster declaring a presentation titled "Water and Climate Justice" to be held at the Museum for Human Rights. I made sure to attend.

One after the other, I heard all but one Indigenous speaker refer to the need to return to the way of life was when they were children or to the life their grandparents led, living off the land and water, how they wanted to drink the water flowing in the streams. I could identify. I wish I could return to the safe space of my childhood, to my grandparents' village, where I grew up.

Their stories, their memories, warmed my heart. I could feel their love for long ago, now displaced. But if I have learned one thing, it is that we can never go back, no matter how much we want to, no matter how hard we try. There is no going back. That is the trouble: once the pollution genie has been let out of the bottle, how do you put it back? Indigenous people have felt the brunt of the damages from all resource development in this country. They have been asking, begging, protesting, demanding, but all their voices have fallen on deaf ears. Because politicians ignored their pleas, and those of many other advocates, the issue has now multiplied exponentially and become dramatically worse today.

Howard Cardinal, an Elder at Saddle Lake First Nation, explained to me that if we take care of the land, the land will take care of us. He also shared that he felt the world is made up of two types of people, Givers and Takers. So true, so simple, so logical.

The panel of speakers at the Water and Climate Justice presentation was led by Elder Sherry Copernace, who has extensive experience in Indigenous social services and embraces her Anishinaabe traditions. She is currently employed at the University of Manitoba in the Master of Social Work based in Indigenous Knowledges program and holds a Masters of Social Work. Recently, she was a member of a team awarded a $2.5 million partnership grant on Water Governance and Indigenous Law by the Social Sciences and Humanities Research Council of Canada (SSHRC).

I feel utter shame that settlers have ruined both the quality and quantity of water across this great country. The focus of the presenters' research was on decolonizing water. It needs to happen. I am encouraged by the many Indigenous groups working together to find a solution, a necessary act of reconciliation. My hope is that one day all our resources are governed by the Indigenous people who are the only true custodians of this land.

Indigenous rights to land are determined by the courts, not by politicians or settler populations. These rights carry legal force independent of governmental or public opinion.

The 1997 Delgamuukw v. British Columbia Supreme Court decision definitively established that "aboriginal title encompasses the right to exclusive use and occupation of the land held pursuant to that title."

This judicial doctrine confirms that Indigenous land rights are not privileges granted by settlers or subject to political whim. Aboriginal title represents inherent ownership of the land itself, including complete authority over land use—rights that predate Canada's formation and persist through legal recognition, beyond the reach of legislative or popular override. The Calder and Delgamuukw decisions are Supreme Court of Canada rulings, which means they apply as legal precedent in all provinces across Canada.

My concern is that our water is now in such a serious state of pollution, it can only be distributed to taps in homes of many First Nations communities if they also embrace ethical, unbiased, conventional science. Their dilemma is knowing who they can trust. Without independent sound science, I question the tangible impact of these treatment systems in improving water quality for communities.

Nicole Wilson, who chaired the presentation, is a scholar of settler origin focusing on Indigenous peoples, environmental governance, and environmental change in the Arctic. Her research examines how Indigenous peoples assert their self-determination and revitalize their governance systems in response to stressors such as climate change and resource development.

All the panel members had diverse and impressive backgrounds. Sameer H. Shar, another presenter, graduated with a Bachelor of Environmental Studies from the University of Waterloo and has experience in developing corporate social and environmental policies in Canada and India. He is part of the EDGES Research Collaborative and the Zerriffi Research Group, focusing on sustainable river basin development in developing countries.

Elder Charlotte Nolin and Colleen James, another presenter, rightly declared that access to water is a human right and emphasized the historical exclusion of Indigenous communities from important water policy decisions. Nolin stressed the critical role of Indigenous communities in shaping future water policy. I agreed wholeheartedly with every word the presenters shared.

After the presenters finished, they invited questions. I asked if they might share their perspective on Bill C-61.

I was surprised to learn that none of them knew anything about Bill C-61. I also asked how their research work contributed to improving the efficacy of water treatment in First Nations communities under DWAs or to stopping industry from releasing toxic effluent into Canada's waterways. They closed the question period without answering my questions.

While they were packing their bags, I gave a copy of my book, *Water Confidential*, to a member of the panel. I called her six months later, but she hadn't read it. I also spoke with another presenter as they were packing up. She asked me if I had any suggestions for her as she had been invited to join the board of the Canada Water Agency (CWA). I told her I had many suggestions. We exchanged phone numbers, and I called her several times for a few months, but she never returned my calls. I can only assume I didn't fit in CWA's approved circle of advisers, which prompted my Freedom of Information (FOI) request to CWA, in an attempt to determine which Indigenous people, or which Indigenous organizations, the CWA had invited to their table. I personally know of two who refused their invitation. You will need to read to the end of this book to learn the answer, it is an incredibly important factor.

I left the event wondering how any of the allotted funding improved drinking water for any First Nations communities in Canada. It's a crucial question we should all ask.

Chapter 16

Freedom of Information (FOI)

FOI requests are a process that should make the government transparent. I was naïve when I began requesting data from different government agencies. I assumed I would receive my requested information. Most of my requests were to ISC; the agency always confirmed receipt of my request and then asked for additional time to comply. More than fifty requests to ISC, over fifteen months, only produced five results.

Ottawa is planning to make access to information even more restrictive: Internal document. The Liberal government is proposing to make the dysfunctional access-to-information system more restrictive, says an internal document.

A discussion paper intended for public release on Oct. 1, but then withheld, proposes the government designate certain individuals as "vexatious" applicants who could be barred from making requests under the Access to Information Act. In my experience this is exactly what has been happening for the past eighteen months, and I assume I must be considered vexatious.

Another proposal would allow federal departments to legally put an information request "on hold" until a requester clarifies to the government's satisfaction precisely what information is being sought.

There's no such "hold" provision in the current legislation. I wish this was in effect, it could have saved numerous hours and dollars for both government departments and requesters.

The paper also suggests diminishing the ability of the information commissioner to make legally binding orders, orders that it claims are overwhelming many departments and leading to lengthy backlogs. This reeks of the opposite of the transparency FOI was intended to provide.

SHOOTING THE MESSENGER

Dr. Irving Selikoff, America's leading asbestos disease expert from the 1960s to 1990s, faced relentless attacks calling him a "media zealot" with "bogus credentials" and "flawed science." Critics labeled him malicious and fraudulent. However, these attacks were orchestrated by the asbestos industry to silence him.

Selikoff exposed how asbestos, which is consumed more heavily in the U.S. than anywhere, caused deadly diseases. Rather than hold corporations accountable for poisoning workers and the public, the industry weaponized doubt against the messenger. This pattern repeats across environmental health fights: when scientists threaten profits by documenting harm, industries attack their credibility rather than address the evidence.

Selikoff's persecution reveals how powerful interests evade accountability by destroying reputations of those who demand they answer for the damage they cause.

Chapter 17

George Gordon, Pasqua, and Yellow Quill First Nations

I was compelled to seek answers after three communities with IBROM water treatment systems were placed on BWAs.

The IBROM system had worked flawlessly for over two decades, and George Gordon, Pasqua, and Yellow Quill First Nations all have highly dedicated WTOs. George Gordon First Nation, a Cree and Anishinaabe community north of Regina, Saskatchewan, received an IBROM system in 2005. Bob Pratt, the WTO, was highly conscientious and founded the Advanced Aboriginal Water Treatment Team. Pasqua First Nation #79, a Nêhiyaw and Anishinaabe community, also received an IBROM system in 2005. Yellow Quill First Nation, a Saulteaux community, was the first to receive an IBROM system in 2004, with Robert Neapetung and later, his daughter Roberta, ensuring safe drinking water for their community.

I questioned why these communities were now on BWAs and found that ISC had not maintained the distribution lines, though the IBROM systems were still functioning perfectly. ISC ignored the questions in my FOI request about maintenance costs and sent me incomplete data, which led to more questions. Each community's expenses more than doubled from 2020 to 2021.

The exact same total expenses were recorded for subsequent years for each community. Some years had no expenses recorded. The figures provided simply were not logical and therefore not believable.

Water Canada published an article in their newsletter for October, the article on per- and polyfluoroalkyl substances (PFAS) stated that "Reverse Osmosis (RO) and nanofiltration (NF) are very effective at removing nearly all PFAS. ... However, these technologies are costly, energy intensive, and create brine or retenate (the solution that is retained by a filter or membrane during a separation process. It is the concentrated portion of the original feed stream that does not pass through the pores of a semi-permeable membrane) that creates disposal challenges. They are most practical for point-of-use and small systems."[44]

I checked with Roberta, the WTO at Yellow Quill First Nation for many years, and she confirmed that after seven years there was still a small amount of biofuel which created a biofilm on the RO, but they didn't know until they dissected the RO, and that despite their higher iron count they found them thriving after seven years.

This despite ISC neglect of maintenance. I don't know what removal rate the IBROM offers for PFAS, but it was not an expensive system as Water Canada stated, and I would say its robust performance is worth every penny.

Eventually, I learned the details were held by the respective First Nations community. I assume this meant that after providing funding, ISC has no checks and balances to account for various expenses? My questions resulted in a copy of a letter from the ISC Capital Manager for Saskatchewan, which appears as Appendix 1. I will also revisit this letter in a later chapter.

Until recently, ISC only funded 80 percent of operations and maintenance of all First Nations infrastructure. I have never figured out where the community was to find the remaining 20 percent. Often, they didn't. Hence maintenance was often nonexistent, as in these three communities. Even the IBROM, as robust as it is, required operational maintenance. When that was not provided, the drinking water quality was compromised, and the communities were put on BWAs.

Susan Blacklin

This simply wouldn't happen in any non-Indigenous communities in Canada.

EDUCATING TOMORROW'S LEADERS

A student from Little Saskatchewan School made history by winning the Manitoba First Nations Science Fair and advancing to the 2017 Canada-Wide Science Fair in Regina. Using water testing kits provided by the Safe Drinking Water Foundation, the student investigated water quality in their own community—the Little Saskatchewan First Nation in Manitoba.

The teacher, Wilson Fallorin, celebrated this achievement, noting how students' enthusiasm for learning about their environment became reality through accessible scientific tools.

The success inspired Wilson to replicate this achievement at his new school, Lawrence Sinclair Memorial School on Kinonjeoshtegon First Nation near Lake Winnipeg, demonstrating how one student's accomplishment could spark continued scientific inquiry within Indigenous communities

Chapter 18

Pikangikum First Nation

None of Pikangikum's homes have running water; the water treatment plant only services a few buildings. This Ojibwe First Nation in Ontario is one of the few Indigenous communities that does not have water distributed to their homes.

Pikangikum First Nation, like Stanley Mission in Saskatchewan, uses a lagoon system for wastewater treatment. Dr. H2O pioneered what I believe to be the first biological wastewater treatment system at Stanley Mission in 2009. I want to know if the released effluent from the sewage treatment at Pikangicum First Nation is upstream or downstream from the intake for the water treatment plant.

Think about treated effluent flowing into a river, and just a few metres downstream are the water intake for the water treatment system. Wouldn't you prefer to take in water upstream, prior to the effluent being discharged?

Pikangikum, with more than 500 homes who have no access to running water for a community of 4,000 people, is suing the Government of Canada in Federal Court, seeking $2 billion in damages and $200 million for urgent repairs at its water treatment plant. Its most recent DWA was issued in February 2024.

Previous advisories were in place from October 2000 to July 2002 and from October 2005 to September 2019. Pikangikum has declared states of emergency in 2000, 2011, and 2015 due to the lack of potable water.[45] The federal government has violated Pikangikum's rights by failing to ensure adequate access to potable water and sewage systems.

During the E. coli crisis in Walkerton, Ontario, in 2000, Pikangikum also had an E.coli outbreak, resulting in one woman needing her arms and legs amputated. The government pumped resources into Walkerton but ignored Pikangikum. The community recently discovered that only one of the three pumps at the pumphouse is working, but it doesn't produce enough pressure. Fixing pumps is at the very bottom of necessary knowledge to maintain safe drinking water. I have had to remove a failing pump from our own primitive water treatment system on the farm and take it to the city to be repaired.

First Nations rarely have the luxury of taking a pump to be fixed – and here lies an opportunity for trades and employment for a person on each reserve. Perhaps it should be incorporated into the job description of WTOs with appropriate training.

ISC spent over $700,000 on a water and wastewater feasibility study and $10.1 million to support the community between 2015 to 2025. Many studies have been done, but no solutions have been implemented. Chief Paddy Peters is correct when he states that in the time it takes to write and review these assessments, his community could have found a solution. Spending $700,000 for a feasibility study? Over $10 million in ten years to support the community? Where has this money gone? This money obviously hasn't improved the water treatment system. I want to see accountability, from ISC, from the community, but above all from those who have benefitted from these contracts. These are our tax dollars that are being thrown into the wind. I tried in vain by filing a FOI request to determine those who benefited from the contracts, as I question whether they fulfilled their obligations ethically.

I submitted a FOI request to find out how the $10 million was spent. ISC demanded an extension of 300 days, claiming they are preparing 31,000 pages of data. I have requested a phone call as I cannot imagine reviewing, let alone compiling 31,000 pages of information. I imagine hundreds of pages of redacted or irrelevant information, and I don't expect a response until April 2026. I filed a complaint, and it was overruled. Apparently 31,000 pages are necessary for such a large request. No one heeded my point that 31,000 pages was a small book and not what I am looking for.

I will post the results on my website when, or if, I receive them. I later learned of a research paper about FOI requests written by Ann Rees as part of her Atkinson Fellowship, Red File Alert: Public Access at Risk, which confirms there are systemic problems with the entire FOI process.

AUTUMN PELTIER: A LONG WALK FOR FIRST NATIONS' WATER RIGHTS

Water shaped Autumn Peltier's life on Manitoulin Island, Lake Huron, where Indigenous communities lacked safe drinking water despite Canada's water abundance.

Her aunt, Josephine Mandamin, founded the "Water Protectors" movement, walking about 25,000 miles around the Great Lakes carrying a copper bucket, reviving women's traditional role as water keepers.

Peltier became a water advocate at the age of eight after witnessing mothers washing babies with bottled water. At 12, she famously confronted Prime Minister Justin Trudeau about pipeline approvals, tearfully criticizing his choices.

The Anishinabek Nation appointed her Chief Water Commissioner at 15. Ironically, she feared deep water after childhood trauma, preferring shorelines despite her water advocacy mission.

Chapter 19

Saddle Lake First Nation

A FOI requesting details of water treatment plant upgrades resulted in an ISC response of over 250 pages. Only thirty-seven pages had information, some of which was irrelevant or duplicate, and all other pages were redacted.

I received various records, most of which were confusing due to being presented in different formats. Annual fuel costs of $1,224 in 2013 jumped to around $10,000 annually in subsequent years. Electricity and gas for the water treatment plant and the pump house amount to between $100,000 and $150,000 annually, despite a solar panel being erected in 2016.

Insurance for buildings runs around $55,000 annually, yet it is noted as a band responsibility, meaning the community is supposed to magically find funds to pay this without ISC funding allocation. Of course, if there is a fire and the community has no insurance, the media would decimate them for being irresponsible. Where are they supposed to find this money for insurance? Yet in later years it is recorded as an ISC expense, reflecting more inconsistencies.

I submitted another FOI request, and the response stated that no records under the control of ISC were related to my request, as the records are under the control of the First Nation.

Under the First Nations Financial Transparency Act, First Nations are required to make their audited consolidated financial statements and remuneration schedules available to members and to post them publicly online.

Although enforcement of the Act was suspended in 2015, many Nations continue to share their statements voluntarily, and ISC still posts them on its website when provided. It appears there is a very fuzzy line between protecting Indigenous sovereignty and providing transparency for how funds are allocated when they do not produce the desired outcome.

ISC/CIRNAC is also required to publish these documents on its website. I find it interesting that the department has not been called Aboriginal Affairs and Northern Development Canada (AANDC) since 2017. Despite this, these audited financial statements are on an AANDC website that was last updated March 18, 2025. That ISC didn't seem to be aware of this, makes me question their competency.

Despite this, the financial statements do not answer questions about why costs like fuel, electricity or insurance jump overnight. I wonder why this colonial terminology is still in use today if the Act has been suspended.

SOLAR BALANCES SUPPLY AND DEMAND

Australia will provide at least three hours of free solar power daily to all households starting in 2026, including those without solar panels. The program targets midday when solar generation peaks, encouraging users to shift electricity consumption to these hours.

Energy Minister Chris Bowen explained that anyone who moves their electricity use to zero-cost periods will benefit directly, regardless of whether they own or rent or have solar panels. Greater participation will create system-wide benefits that lower costs for all users.

Currently, about four million Australian households have rooftop solar panels. During sunny afternoons, solar generation can be so abundant that electricity prices turn negative, while peak demand occurring hours later strains the grid. This initiative aims to balance supply and demand while maximizing renewable energy use.

Chapter 20

Muskoday First Nation

The Water Canada website posted information about the inauguration of a new, advanced water treatment plant on August 15, 2024.[46] This $8.6-million facility, funded by ISC, will deliver safe and reliable water to all residences and key community buildings in Muskoday First Nation.

Chief Ronald Bear expressed pride in this accomplishment, highlighting that the new plant will enable the community to independently manage and treat its own water supply. Chief Bear stated, "This state-of-the-art water treatment plant ensures that the people of Muskoday First Nation have direct access to an abundance of this Sacred Essential Element (Water) to live, thrive, and survive for generations to come."

The facility features advanced biofilters and reverse osmosis technology, ensuring high quality water. It is a key step toward self-determination for the Muskoday First Nation, reducing its reliance on external water sources and enhancing community resilience. It sounds as though the community is benefitting from a treatment system similar to the IBROM which Dr. H_2O pioneered, which piqued my interest. It's not because the IBROM is the only process to be revered or that it is the ultimate system. I hope that it inspires others to continually research, create, and develop optimal systems to provide the highest quality drinking water possible. What I fear, is that cost cutting, or profit maximising imitations of the IBROM system may be built which are not as effective.

I submitted an FOI request on May 5, 2025, keen to learn how effective the system is, and if the system is a duplicate of the IBROM. After the usual delay, ISC requested an additional 120 days. After two FOI requests and two complaints, one of which remains outstanding, I received 821 pages of data, but 773 pages were completely redacted, and only forty-eight pages shared content, many with partial redactions. I cannot tell the specs for the system or who built it.

The data I received did not include the requested analysis of source and treated water, both prior to and upon completion of the new facility.[47] Nor did their response detail the state-of-the-art process, the development of the system or any other communities benefitting from the same system prior to it being implemented at Muskoday First Nation. I am concerned that ISC has not conducted thorough testing of source and treated water, both prior to and following the system becoming operational.

Muskoday First Nation demanded its own water treatment plant, after BWAs were called due to oil spills in the North Saskatchewan River. About 800 band members live on the reserve, which is about fifteen kilometres south of Prince Albert, Saskatchewan. The community is one of several in the area that was connected to a rural water utility normally supplied by Prince Albert's water treatment plant.

I don't know how many citizens depend on the Prince Albert Water Treatment plant for their drinking water. In 2016, many rural communities were thirsting for water after the plant shut down due to an oil spill by Husky Oil.[48]

I cannot find any indication that the Prince Albert water treatment plant has been upgraded. Artificial Intelligence (AI) tells me that the plant serves an estimated population of over 40,000 people in the City of Prince Albert and the surrounding rural areas, including residents within the city limits and approximately 1,000 rural residents in nearby communities who are connected to the plant's infrastructure.

How can we justify spending $8.6 million to provide 800 people with safe drinking water if we still have 40,000 people struggling with unsafe drinking water? Why wasn't the Prince Albert Water Treatment plant upgraded to serve all its customers with the same high quality of drinking water?

FIRST NATION COURT VICTORY

The Yukon Supreme Court quashed government approval for Metallic Minerals Corp.'s ten-year gold and silver exploration project in the Beaver River Watershed, ruling the 2021 decision breached Crown obligations by failing to properly consult the First Nation of Na-Cho Nyak Dun.

Chief Justice Suzanne Duncan returned the proposal to the evaluation stage. The project involved building roads, trails, a helipad, and worker camps. Chief Simon Mervyn hoped the ruling would inspire other First Nations to defend treaty rights and influence land decisions.

He emphasized government must respect treaties through collaborative planning that protects rights, lands and waters. The First Nation thanked Metallic Minerals for suspending exploration during proceedings.

Chapter 21

Grassy Narrows First Nation

You may have heard of the plight of the Grassy Narrows First Nation and its drinking water. Mercury poisoning in their community dates back to the 1960s and '70s, when the Dryden Paper Mill discharged approximately nine tons of mercury into the river system.

In 2020, ISC Minister Marc Miller congratulated Chief Rudy Turtle and the Asubpeechoseewagong Netum Anishinabek (Grassy Narrows First Nation) community for completing upgrades to their water treatment system, which eliminated all long-term DWAs. ISC funded over $5 million of these upgrades, lifting a long-term DWA in effect since June 2014. The improved system now provides clean and safe drinking water.[49]

In 2024, the Grassy Narrows community held a rally at Queen's Park in Toronto to highlight the impact of mercury poisoning in their area. Canadian Broadcasting Corporation's (CBC's) Ali Chiasson reported that the community is seeking compensation for the contamination of the English-Wabigoon River system. The province is leading remediation efforts in the river system, and the federal government says it's spending millions of dollars on a Mercury Care Home, but community members say change isn't happening quickly enough.

Chrissy Isaacs, an activist affected by mercury, questioned why Grassy Narrows has had to fight for over fifty years for clean water and good health, while similar issues in Toronto could be resolved quickly. Grassy Narrows filed a lawsuit against the provincial and federal governments in Ontario's Superior Court of Justice, arguing that the governments violated their duties under Treaty 3 by failing to address mercury contamination in the English-Wabigoon River system. Research from the University of Western Ontario suggests that ongoing industrial pollution from the mill worsens the contamination.[50]

Williamson joined fellow mercury sufferers at the groundbreaking ceremony for the long-awaited Mercury Care Home in Grassy Narrows First Nation. He expressed that this should have happened long ago. A couple of hundred people gathered as a golden shovel hit the ground in the Ojibway community near the Ontario-Manitoba border.

The community has fewer than 1,000 residents, and the care home will provide inpatient services for twenty-two individuals and outpatient services to all affected community members. First promised in 2017 and delayed for several years, the state-of-the-art facility—shaped like a sturgeon—is anticipated to be ready in two to three years.

The federal government says it's spending $82 million on the construction of the 6,500-square-foot home and $68.9 million on community trust to support ongoing operations. It needs to be built; the people of Grassy Narrows deserve far more than a treatment center. But why are Canadian taxpayers on the hook for this expense? It is the Dryden paper mill who caused the pollution. Shouldn't they pay for it? They have reaped the profits from not paying for the damage they caused. But the government allowed them to proceed, year after year, decade after decade. So, is it the company or the government who we should hold accountable, due to their lack of regulations?

I filed another FOI request. I wanted to see the contracts awarded to engineers or consultants. I wanted to see if they were held accountable based on analysis of source and treated waters, if there were deliverables and hold backs until their work was verified effective. I was trying to determine where all this money has gone. I was shocked at ISC's response that no records were available.

I changed my wording, thinking I must not have explained my request adequately, and made a second request. Again, ISC replied that no records existed. I queried possible confusion regarding the name of Grassy Narrows or Asubpeeschoseewagong Anishinabek. ISC confirmed all names were taken into consideration.

Shocked at their responses, I contemplated where I might be wrong. My cell phone rang and a long number flashed on the screen: 111222333444. I ignored it, thinking it was a spam call. Five minutes later the same number called again; this time I answered. A male voice explained: "I see your information request, and I see the reply they sent you. I think I know why."

My mouth must have been gaping open. I explained to him my frustrations at not getting answers to my questions.

"What you have to do is change your words."

"Really? what do I need to ask for then?"

"Don't ask for contracts or agreements, ask instead for grants and contributions."

I thanked him profusely. Almost immediately I began submitting a new request. Again, their response, dated January 15, 2025, indicated that a search of the records under the control of ISC has revealed none to be responsive to the subject of my request. This is impossible!

On June 2, 2016, teenagers from Grassy Narrows travelled 1,700 km to the Ontario legislature to demand action on their community's lack of clean drinking water. They were removed from the chamber for wearing T-shirts with the message "Water is Sacred," which was considered a form of protest.[51] This was considered appalling at the time. Today, we see the USA being dismantled one organization and one law at a time: its democracy eroded, and the rights of women and all minorities disappearing before our eyes. I see this eviction of the students like a Trump project of DOGE or ICE, or a MAGA threat. All acronyms we have become far too familiar with.[52]

I have a right to request copies of these contracts and expect to receive them with no further delays. $82 million for construction of the care home, $68.9 million for a community trust fund, and $5 million for upgrades to the water treatment plant which still isn't producing safe drinking water. A total of over $156 million, for 1,000 people. What I want to know is if the contracts awarded to third parties to upgrade the water treatment plant state measurables, deliverables, or other means to establish that effective systems were built. I doubt it. Is that the fault of ISC or the First Nations or the contractors?

I submitted yet another request, this time citing the numerous government statements, each confirming ISC public declarations of funding to Grassy Narrows. While this has a bad smell, possibly of corruption, don't assume the First Nations leaders are the culprits. I am leaning toward those receiving the contracts as being the perpetrators. Who deposited these millions of dollars and left the community without safe drinking water? A conversation I had with a civil servant explained a lot to me. ISC is presently receiving an increase of 75 percent for requests for information.

Is it that there is no leadership, no direction, no operating procedures, or maybe they simply don't track the information I am requesting? Months passed and I received a response asking me to limit my request, after numerous FOI requests and complaints resulted in no records available, the ISC has now found 37,500 pages of information! I will post the information on my website, if I ever receive it.

There are many First Nations communities all over Canada that are struggling with the impact of industrial waste in their source waters. In January 2026, ISC Minister Mandy Gull-Masty's refusal to commit to source water protections in a promised clean water bill shows the government is sidelining the health of Indigenous communities in its push to build up the economy. I call it genocide. If you don't find this a problem, do you think polluted source waters only affect Indigenous communities? There are no barriers to polluted source waters. They affect every Canadian.

On July 4, 2025, First Nations communities came together with allies to say "no" to the proposed nuclear storage facility in Treaty 3 (Northwestern Ontario). Grassy Narrows has voiced concern over the project, planned near the headwaters that feed their territory.

A proposed Kinross gold mine near Red Lake, Ontario, has raised significant concerns among the Grassy Narrows First Nation due to its potential impact on the nearby river system, which is a vital resource for the community. The mine, if developed, would discharge treated wastewater into the Chukuni River, a tributary of the Wabigoon River, which was previously contaminated by mercury from the Dryden pulp mill. This has led to fears that the mine's discharges could worsen the existing mercury contamination and negatively affect the community's health and traditional way of life.

Earlier in 2025, a provincial tribunal struck down one of Kinross' permits after a legal challenge by Grassy Narrows. The tribunal found "good reason to believe no reasonable person" could have approved the permit, calling it based on "incomplete" studies and warning of a "threat of serious environmental impacts." Kinross withdrew that permit but has already applied for a new one.[53]

The *Toronto Star* also reported July 10, 2025, that Kinross' USA. subsidiary, Crown Resources, admitted to over 3,500 violations of its water permit at the Buckhorn Mountain mine in Washington State.[54] The Washington Attorney General is now suing the company under the Clean Water Act. Environmental groups say much of this could have been avoided if action had been taken years ago.

Scientists found that continued discharge of wastewater from the Dryden Paper Mill, combined with existing mercury, has created high levels of methylmercury, which is an even more toxic compound.[55]

"Other forms of mercury don't accumulate as strongly as methylmercury, but because it accumulates, it builds up to high levels in organisms, presenting that greater risk," said Brian Branfireun, a biology professor at the University of Western Ontario. "It's actually more serious than I even imagined."

Kinross' latest application comes as Ontario advances Bill 5, a controversial law aimed at fast-tracking mining approvals. Grassy Narrows fears the province will use it to silence opposition and rubber-stamp dangerous projects.

Kinross projects have faced Indigenous and environmental resistance from Washington State to Alaska. The Native Village of Dot Lake has initiated a groundbreaking lawsuit against the US Army

Corps of Engineers, challenging the authorization of Alaska's largest suction dredge mining operation, a move that underscores the ongoing struggle between Indigenous rights and commercial development interests.[56] The proposed mining project, spearheaded by Kinross Gold Corp. and Peak Gold LLC, is set to transform the landscape near Tok, Alaska, by establishing the state's largest suction dredge gold mining operation in an untouched estuary.

These corporations frame their projects as beneficial, but the long-term risks to health, culture, and future generations cannot be undone. How can a community of 1,000 people possibly defend so many assaults on its people?

EVERYBODY IS A FIGHTER IN GRASSY

December 2025 marks twenty-three years since Grassy Narrows began one of Canada's longest blockades against industrial logging on their land. Despite enduring intergenerational mercury poisoning, the community has maintained steadfast resistance since 2002.

Their persistence achieved a historic logging moratorium, recently extended through 2033, to safeguard their territory. "Everybody now is a fighter in Grassy. Nobody wants logging or mining. The more we fight, the more we gain," says JB Fobister, who leads Grassy Narrows' Land Protection Team. Their ongoing struggle demonstrates remarkable resilience in protecting Indigenous lands.

"Seven x seven generations: Everything we do now will impact our grandchildren for seven generations. If we discontinue our negligence, we can change things around." - Bidassige-Ba/Josephine-Ba Mandamin Ogimaa-Kwe Water Walker

Chapter 22

Neskantaga First Nation

In Ontario, the Neskantaga First Nation community of 374 people is working to upgrade its water treatment system to end a long-term DWA. This advisory, affecting seventy-six households and six community buildings, was issued in February 1995 and became long-term in February 1996, making its thirty-year BWA the longest BWA in Canada.

The water treatment system's construction was mostly completed and commissioned in December 2020 but some issues remain. Moonias challenged the government to provide a water operator in Neskantaga several years ago. As a result, the Ontario Clean Water Agency has been running the plant since 2020, with support from ISC.[57]

ISC is funding the Ontario Clean Water Agency to oversee the water system and train local operators. They are also supporting a Water Systems Infrastructure Assessment to ensure the system meets long-term community needs. ISC continues to work with Neskantaga First Nation to build trust in the community's water by funding the "Trust the Taps" initiative, which includes a community healing plan and mental wellness programs.

I began submitting FOI requests for Neskantaga First Nation in

July 2024. My goal was to review all contracts awarded by ISC, to determine if they contain analysis of source and treated waters both prior to and following any upgrades to determine the efficacy of the system. I was told no records were available, so I sent another FOI request. They replied asking for keywords, then requested additional time and finally, once again, I was told no information was available.

Ottawa has spent $30 million on upgrades to Neskantaga First Nation's water treatment plant. Recently, Chief Moonias requested $52 million from Ottawa for a new plant, and ISC Minister Patty Hajdu supports this plan. Apparently. Despite recent upgrades, the plant's distribution system is flawed. Talks have been underway for way too long on designing a new plant with a new water intake to provide cleaner source water.[58]

I am suspicious that previous water treatment plant systems may have been built in compromising locations, perhaps downstream of treated sewage or, worse, downstream of ineffectively treated sewage. Or that the systems contracted were not the best system for the challenges in the source water. If any of these scenarios, or others, exist, then I want to determine responsibility for such negligence. I submitted another FOI, citing all the media announcements confirming ISC funding.

Chief Moonias said the plant is producing clean water now, but there are problems with the distribution to people's homes. This could be because families in many First Nations communities frequently leave their homes to live in the bush for weeks or months at a time, causing water to sit and become stagnant in distribution lines. Without regular use or flushing, systems become stagnant and it is costly to remove bacteria from stagnant lines. It is almost impossible to maintain a safe standard of drinking water in distribution pipes that are not used every day of the year.[59]

In Neskantaga's case, Moonias describes a patchwork of short-term solutions over the years that have cost tens of millions of dollars and haven't addressed the root of the problem. He said the lack of access to clean drinking water has taken a toll on his people's mental health and physical conditions like skin rashes. Community members, including the First Nation's health director, have long linked skin rashes with people showering with contaminated water.

Moonias said he submitted a project approval request to Ottawa in February 2025, when the government was prorogued, in hopes of getting funding for a new water treatment plant, at an estimated cost of $52 million. This figure seems pie in the sky to me.

I have tried contacting Chief Moonias directly for many months. Their office's phone lines are down. Their website has no email addresses, no contact information. Is this the result of their need for independence and privacy? Or does it indicate that more than their drinking water is in crisis? In 2019, Chief Moonias ceased all operations on the new water treatment plant and ordered workers and officials from Kingdom Construction Limited to "leave the community immediately." At the same time, Moonias also asked ISC to decommission the temporary water filter system and supply the community with unlimited bottles of water. He explained that the only safe, clean water that residents in Neskantaga had access to was from the temporary RO unit, which is in a small wooden shack near Attawapiskat Lake. Moonias said they had been using the temporary filter system for about ten years and people in the community "are getting tired of using the RO Unit."

Ten years of no access to safe drinking water is absolutely deplorable. I try to imagine the mental toll it must take as well as the physical illnesses. However, decommissioning a temporary treatment system which provided drinking water in bottles to the community, only to replace it with bottled water that was flown in, is also ridiculous. That is assuming their bottled water was tested regularly and met all guidelines.

I learned that the reverse osmosis unit, located inside this wooden shack, often has problems as it is temperature sensitive and not meant to be operating out in the cold. I want to know who built this temporary system. They should have known exactly how cold it gets there in the winter. This is why I have filed so many FOIs; I want to know who has been paid—and I imagine paid very well—to build ineffective systems such as this. With $30 million being shared with engineers and consultants over thirty years, someone needs to be held accountable.

For Neskantaga alone, I filed four FOIs. All four FOIs were futile. Three stated no records available. Three complaints were ruled to be ineligible. Then, from ISC, came a reply to my fourth complaint. This reply was more detailed than any previous communication: "The system does not include records related to third parties hired directly by First Nations using ISC funding. While agreements with third parties for water treatment may exist within Neskantaga First Nation, such records would not be under the control of ISC if they were negotiated directly by the First Nations using ISC funding."

I was shocked. ISC is writing cheques for millions of dollars with no ability to ensure the funds are spent wisely. To my way of thinking, ISC, who have repeatedly failed to provide Neskantaga First Nation and many other First Nations communities with safe drinking water, has an ethical and moral responsibility to guard against anyone taking advantage of the community. I can also appreciate the need for Indigenous communities to be independent, to respect their privacy and support self-determination.

What about the individuals and/or companies who received $30 million in the past, and failed to fulfill their obligations to provide an effective water treatment system? Some people are overly eager to make money, not caring about those they abuse in the process. I emailed Neskantag's new Chief elected spring 2025, Chief Gary Quisess, and left a phone message. I made it clear that I suspect those contracted have not lived up to their responsibility, but he did not return my call or email.

Ontario and Alberta signed memorandums of understanding the summer of 2025 to explore building a railway to connect northern mines to processing facilities in the south. While the governments want to advance development of resources, "This is a country where First Nations people agreed to share the land with the settlers. The way we're treated today is not right, it's a hidden genocide," Quisess said.[60]

Paul Dubé, the Ontario Ombudsman, toured Neskantaga First Nation at the invitation of Chief Gary Quisses, and called on the provincial and federal governments "to take immediate action to address the unacceptable and unsafe conditions" in the community.

"I'm very unsettled by this visit and we want to find ways to contribute our services," Dubé told CBC News.

I tried contacting the office of the Ontario Ombudsman. They must experience a lot of complaints as the first direction on their phone message system is for how to make a complaint. Three times I called and asked how I could send an email as their given email addresses kicked back. When a real person eventually returned my call, they thought I was filing a complaint. I suggested that they scrutinize the contracts that have been paid and determine if they delivered the services for which they were contracted.

A woman from their office called me October 24, 2025 and explained that the Ontario Ombudsman has no authority over First Nations as they are under federal jurisdiction. I asked her why he had visited the community, on taxpayer dollars no doubt, if there was nothing he could do. She hastily responded that it was to support reconciliation. I suggested that providing them with safe drinking water would be a true act of reconciliation. I asked her to reply to me via email not over the phone. Four days later I received this letter, in which she states: "Our work aims to support the special constitutional relationship that Indigenous peoples hold, inform constructive government-to-government dialogue, and contribute meaningfully to Ontario's reconciliation journey."

The letter offers a link to their services, which states: "If you are Indigenous, we will make sure your rights and cultural safety are respected." If it's not their jurisdiction, I wonder how they can uphold such statements. It's all very confusing to me.[61]

My concern remains where has the $30 million gone that has already been given to the community. I submitted four FOIs over sixteen months, and only the fourth FOI eventually produced limited, mostly redacted, results just days before publishing this book. The results and a summary will be available on my website.

A WOMAN'S JOB

Forty years ago, 90 percent of Icelandic women went on strike, refusing to work, cook, or care for children. This single day transformed Iceland, launching it toward becoming "the world's most feminist country."

The strike's impact was so profound that when Vigdis Finnbogadottir became Europe's first female president in 1980, Icelandic children assumed leadership was a woman's job. One boy protested Ronald Reagan's presidency: "He can't be president—he's a man!" Vigdis served sixteen years and credits the 1975 strike for making her presidency possible.

The lesson: Collective action by women demanding accountability created systemic change. When women refuse to be invisible, they reshape what's possible for everyone.

Chapter 23

First Nations in British Columbia

A reader of my first book informed me about a company that was reportedly installing advanced water treatment systems in First Nations communities in British Columbia.

I reached out to the company, but initially they were unresponsive. I called in unexpectedly at their place of business, and the owner gave me a tour of his business, which is when I discovered that none of its staff had a science background. This raised doubts for me as to its ability to build effective water treatment plants.

Four FOI requests and four complaints to this BC company and none yielded any results from ISC. The saga of no cooperation continued. I wanted to find out if the ISC had any holdbacks to ensure the system's effectiveness and what tests were conducted before payment was made. My fifth FOI replied they needed an additional 120 days, as they were collating over 9,905 pages. I filed a complaint when once again no information was forthcoming.

I spoke with a WTO in the same geographical area, who explained that the water was so pristine it rarely required treatment. This reminded me of Dr. H$_2$O's claim that few communities in British Columbia treated their water, which might correlate with the high number of DWAs in British Columbia. As of October 7, 2025, British

Columbia had 513 DWAs, the highest in Canada, more than double the number of DWAs in Saskatchewan, which had the second most, or triple the number in effect in most other provinces.

If the source waters are not treated or protected, and ineffective systems continue to be built without verifying their efficacy, British Columbia might soon face a crisis similar to that which took place in Walkerton.

Chapter 24

Reconciliation

In 1970, I was leaving England to live in Canada, and travelled to Canada House for the required interview to complete my application to emigrate. I had no idea of the deceit as we were shown videos of never-ending wheat fields, glistening and dancing in a soft breeze; Royal Canadian Mounted Police (RCMP) in their official dress of red serge, standing on every street corner; Indigenous people dancing in full regalia. I'm not sure which is most shameful: how they depicted and "sold" Canada based on the RCMP and the Indigenous people, or how they hid the fact that residential schools were still operating in many provinces.

My first full-time job was at Eaton's. I earned $1.30 an hour, worked a forty-hour work week, with two weeks' vacation after one year. I thought I had arrived in the land of opportunity! Men, of course, in the same position earned $2 an hour or more. Still, the life I dreamed of was within reach. I earned in a day more than I earned in a week in England. A savings account earned eight percent interest. A townhouse sold for $18,000, the equivalent of seven years' salary at minimum wage. I was in my glory. I had no idea back then of what was happening, hidden behind the life I was enjoying.

In the late 1970s, the top 1 percent of income earners held

approximately 8 percent of the total income share. This share rose sharply and peaked at nearly 14 percent in 2007, and by the late 2000s, had doubled to around 16 percent. In contrast, the income for the bottom 70 percent of the distribution was worse off in the late 2010s than it would have been under a 1970s distribution.

Canada's infant mortality rates have improved over time, from about 10.9 deaths out of 1,000 births in 1980 to 4.4 deaths per 1,000 births in 2021, but the country has slipped in its ranking of infant mortality among wealthy, developed nations in 2021.[62]

There are significant disparities in infant mortality rates in Canada. Infants in the most materially deprived areas die at 1.6 times the rate of those in the least deprived areas. The disparities are even more pronounced in areas with higher Indigenous populations, with Inuit communities experiencing 3.9 times higher mortality rates, First Nations 2.3 times higher, and Métis 1.9 times higher. These Indigenous inequities can be attributed to the historical and ongoing impacts of colonialism, including forced relocation, land dispossession, the reserve system, suppression of languages and cultures, residential schools, and unaddressed intergenerational trauma. While unsafe drinking water may be a factor in infant deaths, not all infant deaths can be attributed to unsafe drinking water. However, access to safe drinking water could be the foundation of improving all other inequities.

Significant health gaps exist between the Indigenous and non-Indigenous populations in Canada. To identify and close these gaps, the Truth and Reconciliation Commission of Canada recommendation number nineteen (of the 94 Calls to Action) called upon the federal government to publish data and assess long-term trends for a number of health indicators, including life expectancy among First Nations people, Métis, and Inuit. Estimating the life expectancy of the Indigenous population is methodologically challenging since death registrations do not usually collect information on whether the deceased was Indigenous. It is also complicated by incomplete enumeration of many Indigenous communities.

Authors Eva Jewell and Ian Mosby published "Calls to Action

Susan Blacklin

Accountability." In their words, "In the short time we have been annually observing Canada's record on its supposed progress, we've held the tension of the promise of reconciliation with the actual reality —exacerbated by the deep chasm between the two."[63]

Eight years since the release of the 94 Calls to Action, 81 calls remain unfulfilled.[64] I'm not sure which is most shameful, the lack of due diligence to reconciliation or the lack of due diligence to provide Canadians with safe drinking water.

Globally, billionaire wealth grew by $2.8 trillion in 2024 alone, equivalent to roughly $7.9 billion a day, which has risen three times faster in 2024 than 2023. Meanwhile, the number of people living in poverty—nearly 3.6 billion people—has barely changed since 1990. In Canada, total billionaire wealth stood at $496.76 billion with 65 billionaires. In 2024, billionaire wealth increased by $309 million per day while 3.8 million people live below the poverty line in Canada. Canadian billionaire wealth could easily carpet most of Vancouver (93 percent) in CAN $50 notes.[65]

Imagine (my favourite song) if here in Canada, as in Switzerland, we could oppose moves proposed by our government. Imagine if reconciliation was real. Imagine if all our voices could be heard.

BALKAN RIVER DEFENCE

With over 2,700 dams proposed for Europe's last free-flowing rivers, the Balkan Rivers Tour launched in 2015 as a rebellious effort to paddle these rivers and showcase what damming for hydroelectric projects would destroy.

It spawned Europe's largest direct-action river conservation movement. In one expedition, a four-kayaker Balkan River Defence crew paddled Slovenia's entire Sava River in eleven days, collecting environmental DNA samples and conducting the first complete nesting-season water bird census to oppose ten proposed dams.

Core team member Carmen Kuntz reported they premiered the documentary *One for the River: The Sava Story* and released a paddler's guidebook. The organization aimed to shift public perception toward energy alternatives.

Part Three

Friends of Government

Chapter 25

Does Taxpayer-Funded Research Actually Serve the Public Good?

After months of searching for any sign of biological water treatment processes being pursued by scientists after Dr. H_2O initiated the concept in 2004, I was ecstatic to find this research project, titled "Reducing Pollutants the Biological Way."

A collaboration between the University of Toronto and Geosyntec Consultants Inc. has significantly improved the process of cleaning up common and persistent groundwater pollutants. This partnership, which began over a decade ago and is led by chemical engineering professor Elizabeth Edwards, focuses on using bacteria to detoxify underground pollutants. Their most notable achievement involves a microbial culture called KB-1, which transforms a common dry-cleaning agent and related industrial solvents into harmless by-products. These chemicals have been in use for over eighty years and are associated with serious health issues. The licensed technology has already been successfully implemented at more than 200 sites in the United States and Europe. In recognition of their work, the partnership received a $200,000 Natural Sciences and Engineering Research Council of Canada (NSERC) Synergy Award for Innovation in 2009. Additionally, Dr. Edwards independently received a Discovery

Accelerator Supplement, increasing her five-year discovery research funding to $395,000.[66]

I would like to know if Elizabeth Edwards research has been applied to benefit any communities in Canada. For over eighty years a dry-cleaning agent has been a known cause of serious health issues! My question is why has it taken so long? Why wasn't it banned decades ago? And has KB-1 been tested or, better still, applied in Canada yet? What are we waiting for? A third research project funded by NSERC also gives me hope for the future.

Ensuring the safety of water from its source to the tap is crucial for providing communities with clean and safe drinking water. Manuel Rodriguez from Université Laval investigates the occurrence of disinfection by-products in drinking water over time and space.[67] These by-products are formed when disinfectants used to treat drinking water react with natural substances in the water. His research aims to enhance the understanding of water quality variability at the source.[68] Specifically, he examines how the quality of source water changes over time, with the goal of developing early warning systems to help water managers make timely decisions. In 2010, Dr. Rodriguez received $380,000 in support from the NSERC Discovery Grant and Discovery Accelerator Supplement, which funded his research through 2015. I wonder if his research has been applied in any communities in Canada.

Charities continually attempt to find a sponsor/donor who makes funds available to pursue the issues which they know are prevalent. Too often potential donors have their own agendas, thereby limiting charity's ability to pursue what they know is necessary. I wonder, do scientists also strive to find funders to enable their research of what they know is prevalent? I know that was always Dr. H_2O's dilemma. Too many funders, often government agencies, had their own focus held above the desire of the scientists. When scientific research is funded, like any of these projects, my big question is when or where is it applied?

The federal government funds numerous research organizations, and subsequently thousands of scientists and projects, but serious questions remain about whether this investment truly benefits Canadians or just enriches corporations.

I distinctly recall conversations with Dr. H_2O and his fellow scientists who expressed deep frustration on two fronts. First, they struggled repeatedly to secure funding for research into improved water treatment processes, the kind of practical work that could directly protect public health. Their applications were often rejected. Second, when they did manage to obtain funding, their work led to publications in highly acclaimed scientific journals, generating academic prestige, yet these findings rarely translated into real-world applications that could benefit Canadians. They felt certain individuals financially benefitted rather than the greater good being served.

These concerns were raised twenty to thirty years ago. I wanted to find out if the same problems persist today, with taxpayer-funded research still languishing in academic journals while corporate agendas and profiteering prevail.

Research priorities appear focused on what generates revenue rather than what protects public health. Studies analyzing contaminants in source water or health issues from unsafe drinking water receive funding, while I found little in the way of improved water treatment processes, pollution prevention, and practical solutions, all of which appear to remain underfunded and/or the science rarely applied. Research into water purification for remote communities struggles for funding while billions flow into satellite technology, artificial intelligence or more efficient fossil fuel extraction.

Capitalism treats knowledge as a commodity to be owned and sold for profit, while science ideally pursues truth and serves humanity. These goals often conflict. Capitalism asks, "Will this make money?" while science should ask "Will this help humanity?" When profit determines what gets researched and how findings are used, or not, society loses potentially life-saving discoveries that aren't commercially viable.

Improving public water safety serves everyone but enriches no one specifically, so it struggles for funding. The question isn't whether capitalism influences research—it absolutely does. The question is whether we're comfortable with profit motives determining what knowledge we pursue and who benefits from it.

I researched and submitted an FOI request to the National Research Council (NRC), NSERC, Canadian Institutes of Health Research, SSHRC, Canada Water Network (CWN), Brace Water Centre, and the Water Canada publication. From all of these organizations, I was unable to find research projects that benefited communities, either with improved effectiveness of water treatment systems or effectively removing pollutants from source waters. Neither did any projects I explored suggest that improved national drinking water regulations might bring Canada up to par with European and other developed countries.

A good example is the University of Saskatchewan's Global Institute for Water Security which conducts research on Saskatchewan's water resources. The $1.65 million *Food-Water Nexus Education and Training* program trains graduate students, while the NSERC's Collaborative Research and Training Experience program addresses water quality and human health, with funding of $77.8 million.[69]

Saskatchewan has some of the most challenging source waters to treat. I would love to find just one community which has improved quality of drinking water from these millions of dollars in funding. Or one improved drinking water treatment process or one step toward protecting source waters. I wonder what good does that funding do? Who benefits?

Billions of dollars are being shared annually with a host of scientific researchers. It is challenging to determine an exact figure as funds are often redistributed between organizations. The same dollar could be counted more than once. I imagine each organisation who handled the money took their commission off the top. All I see is good money being wasted. Why fund and conduct research if you are not going to implement its results?

I see a strong similarity between how ISC provides tax payer money to many First Nations communities yet has no means to verify if the funds result in effective treatment systems, to how research organizations share taxpayer money and have no accounting for how effective or, more importantly, how any of their projects are applied for the greater good. Often research projects are based on AI or satellites or on developing equipment that has a high probability of making someone wealthy.

I would like to share with you the organization I found to be the tardiest, the least trustworthy, how over sixteen months they continued to respond to an FOI and still make empty promises to provide the information "in a few weeks." The CWA takes first place.

CROWS ARE SMARTER THAN AI

Swedish startup Corvid Cleaning trains wild crows in Södertälje to collect cigarette butts using a vending machine that rewards them with food. The crows' remarkable intelligence—comparable to a seven-year-old child—enables them to learn this task through operant conditioning and teach each other, showcasing cognitive abilities that surpass simple AI programming.

Unlike artificial intelligence that requires constant human programming and maintenance, these birds voluntarily problem-solve, adapt, and collaborate independently.

The initiative reduces municipal cleaning costs while targeting cigarette filters, a major plastic pollutant. This brilliant natural intelligence demonstrates that crows can accomplish complex environmental tasks through innate learning and social cooperation that even sophisticated AI systems struggle to replicate autonomously.

Chapter 26

Canada Water Agency (CWA)

The CWA was initially established as a branch within Environment and Climate Change Canada (ECCC) in June 2023. It officially became a stand-alone, independent agency on October 15, 2024, when the legislation creating it, part of Bill C-59, received royal assent and came into force. The federal government has invested $85.1 million for the creation of the CWA as well as $650 million for the first ten years for the Freshwater Ecosystem Initiatives (FEIs).

FEIs is a federal program in Canada that supports projects to protect and restore the health of nationally significant freshwater bodies, such as the Great Lakes, Lake Winnipeg, and the Fraser River. The CWA website states: "The Canada Water Agency works with provinces, territories, Indigenous peoples, local authorities, scientists and others to find the best ways to keep our water safe, clean and well-managed for future generations. The overall goal is to improve water quality and aquatic ecosystems for the benefit of communities and the environment."

I know of one Indigenous organization and one Indigenous person whom the CWA had invited to collaborate. The invitees felt they were a last-minute thought, just like when the government made press announcements and would put a token Indigenous person in direct sight of the cameras. They both declined their invitations. I felt the CWA appeared to be looking for token First Nations to verify they were incorporating Indigenous values into their programs.

I filed two requests for information in the summer of 2024. I wanted to understand the agency's total budget and its expenses, especially the salaries it was paying its top executives. Most of all I was interested in learning about its outreach to First Nations and other nonprofit stakeholders to collaborate, and the subsequent responses of those individuals or organizations. I also requested communication, both internally within each of these organizations, and between the ECCC with the CWA pertaining to selection of First Nations individuals, communities, organizations to invite to collaborate or consult with the CWA. After almost a year, I was back where I started because the agency presented me with excessive excuses and then ghosted me for months. The complaints I filed were ruled unjustified.

Imagine my surprise when nine months later, April 2025, I was contacted by someone working at the CWA to provide information on my first request. His explanation for the almost ten-month delay: the CWA was part of ECCC until October 2024 (prior to me making these two requests) but it is now its own stand-alone agency with its own Access to Information Office.

My contact demanded that I cancel the subsequent FOI I had submitted, assuring me he would release the information requested in the next week or so. He did not keep his part of the bargain! Furthermore, he asked that I accept modifications to my request, and additionally requested that I exclude documents that contain possible Cabinet Confidences, as that allowed them to avoid having to consult with ECCC Legal Services.

I had to ask what comprised Cabinet Confidences, to which I learned that in order to reach final decisions, ministers must be able to express their views freely during the discussions held in Cabinet. To allow the exchange of views to be disclosed publicly would result in the erosion of the collective responsibility of ministers. As a result, the collective decision-making process has traditionally been protected by the rule of confidentiality, which upholds the principle of collective responsibility and enables ministers to engage in full and frank discussions necessary for the effective functioning of a Cabinet system.[70]

The Supreme Court of Canada has recognized that Cabinet confidentiality is essential to good government. In the decision Babcock v. Canada (Attorney General), 2002 SCC 57 at paragraph 18, the Court explained the reasons for this: "The process of democratic governance works best when Cabinet members charged with government policy and decision-making are free to express themselves around the Cabinet table unreservedly."[71]

I disagree with the Supreme Court of Canada. I believe democratic governance works best when cabinet ministers charged with government policy and decision making are free to express themselves around the table unreservedly—and that Canadians who elected those individuals have a right to know how the individuals they elected act, and what positions they take on all issues. Surely elected officials first responsibility is to their constituents, not to insidious agendas hiding behind closed doors or to voting in accordance with their respective party leaders. Where is the line between a party leader establishing a strategy or plan forward, and becoming a dictator?

It is easy to forget our rights as politics have become top down, with a "vote with the party agenda or else" implied slogan. How much higher than the Supreme Court do we have to go to find accountability? Or is the Supreme Court just another level of out-of-touch judges who renege on their responsibilities to Canadians? Andrew Coyne reminds me, in *The Crisis of Canadian Democracy*, that government in a parliamentary democracy is supposed to be answerable to the people through their elected representatives in Parliament.

The FOI representative promised he would send me my information within ninety days, but the September deadline passed once again, and it was another "cheque's in the mail" excuse. By November 2025, the information still had not been received. So, I wait. I also continue to search for projects where science funded by CWA has been applied to benefit Canadians.

"Canada's fresh water is threatened by human activity, invasive species, and climate change. Clean fresh water is essential to the economy, the environment, and the well -being of Canadians. Protecting fresh water is crucial." This headline from the Honourable Jean-Yves Duclos, Minister of Public Services and Procurement, on behalf of the Honourable Steven Guilbeault, Minister of Environment and Climate Change and Minister responsible for the CWA, announcing further funding.

We don't need to be told that Canadians are committed to protecting source waters; we already know that. But we don't have a voice. The government chooses to listen to lobbyists and ignore Canadians. What I cannot find is any politician, or any government action, committed to, and acting to protect, our source waters.

"Collaboration between local communities, Indigenous peoples, and youth is at the heart of freshwater protection. Together, we are developing innovative solutions to environmental challenges in Québec and Canada."[72] The last time I checked, Quebéc was a province in Canada, just like all the other provinces. I wonder why they need to be mentioned specifically? How are local communities collaborating? Indigenous people have been begging for decades; now they are demanding that they have safe drinking water. What innovative solutions have been funded and applied to provide them and all other communities in Canada with safe drinking water?

A media release told me that on February 14, 2025, Canada announced over $5.5 million in funding for sixty-five freshwater protection projects through the EcoAction Community Funding Program and Community Interaction Program.

Notable recipients received $100,000 for rainwater collection near Québec City; $61,199 to reduce lake contamination; $95,140 for wetland ecosystem restoration; $200,000 for agricultural buffer strips; $100,000 for Indigenous youth education and $324,349 for turtle habitat restoration biodiversity improvements.

In 2020, mining was responsible for over half (54 percent) of all reported industrial pollutant releases in Canada, with significant impacts on source waters through tailing management and potential leakage. The paper industry, food manufacturing, and chemical manufacturing are top contributors, often discharging into surface freshwater bodies. We need national enforceable laws that require all industries to clean up their effluent before releasing it into our waters.

I cannot help but ask: are these the best ways to keep our water safe, clean and well-managed for future generations? Why is there such a heavy emphasis on funding allocated to organizations in Quebec? How do any of these projects stop pollution of our source waters? Why keep assessing what is polluting our source waters? Or how much pollution is added where? And then paying to apply band-aid solutions? We must stop the pollution!

On October 21, 2025, sixteen months after I filed the FOI request, CWA responded with budget costs for the CWA. Sixteen months to wait for sixteen pages, to learn that top scientist salaries range between $134,000 and $228,000 annually. Their annual operating budget ranges from $73,138,678 for year-end 2026, to $80,397,660 for year-end 2030 for a total of $384,683,042 over five years.[73] Three hundred and eighty-four million, six hundred eighty-three thousand, forty-two dollars. And I don't see one cent of this money improving any community drinking water or protecting any source waters from increasing pollution.

THE WOVEN PROJECT

Indigenous environmental leader Eriel Deranger, from Athabasca Chipewyan First Nation in northern Alberta, established a global Indigenous advisory council on climate change after winning the $4 million Climate Breakthrough Award in 2024.

The council comprises Indigenous climate leaders from seven sociocultural regions worldwide and is the first focused exclusively on climate justice.

Living downstream from Alberta's tar sands, Deranger has advocated for climate justice for nearly two decades. The initiative addresses gaps in Indigenous collaboration on global climate solutions by honoring distinct Indigenous cultures and regional successes rather than simply merging Western and Indigenous knowledge. It develops climate solutions rooted in Indigenous knowledge and rights-based frameworks aimed at achieving land back.

Chapter 27

Big Oil's Global Assault

I am touched that many First Nations people and organizations in Canada express strong solidarity and sympathy with Palestinians, seeing parallels in struggles against colonialism, land dispossession, and resource extraction, with groups like the Assembly of First Nations supporting Palestinian rights.

I see strong similarities between Israel withholding water from Palestinians and Canada withholding safe drinking water from First Nation communities. I call withholding safe drinking water, genocide, no matter where it happens. This recent report from Amnesty International gives me the fuel to continue bringing these atrocities to readers attention.

A first-of-its-kind global mapping exercise by Amnesty International and Better Planet Laboratory at the University of Colorado Boulder has exposed the staggering scale of harm inflicted by the fossil fuel industry on Indigenous Peoples worldwide.[74] This groundbreaking investigation, detailed in the report *Extraction Extinction: Why the lifecycle of fossil fuels threatens life, nature, and human rights*, reveals how Big Oil has systematically colonized Indigenous territories, forcing native communities to bear a grossly disproportionate burden of environmental destruction, health crises, and cultural genocide.

The research demonstrates that at least two billion people globally —roughly a quarter of the world's population—face significant risks from living within five kilometres of more than 18,000 operating fossil fuel infrastructure sites distributed across 170 countries. Among this endangered population, 520 million are children, and at least 463 million people live within just one kilometre of these toxic sites, exposing them to exponentially higher environmental and health risks.

Indigenous lands bear disproportionate fossil fuel infrastructure burden: over 16 percent of global sites deliberately target Indigenous territories, exemplifying environmental racism and colonial exploitation. At least 32 percent overlap critical ecosystems Indigenous communities have protected for generations. Despite climate pledges, 3,500 additional sites are proposed, threatening 135 million more people. This systematic expansion into vital ecosystems directly contradicts governments' stated climate commitments while perpetuating injustice.

The report draws on in-depth qualitative research conducted in partnership with Columbia Law School's Smith Family Human Rights Clinic, consisting of interviews with more than ninety people, including Indigenous land defenders who face existential threats from Big Oil's expansion.

Canada's Wet'suwet'en land defenders face "intergenerational battle fatigue" resisting Canada's Coastal GasLink pipeline, experiencing state-sanctioned violence while protecting their territory. Civil injunctions criminalize Indigenous resistance, enabling militarization and corporate extraction without consent. Defenders emphasize they "were never the instigators" yet bear violence's brunt. New compressor construction threatens Wet'suwet'en survival, exemplifying how colonial legal mechanisms facilitate destructive extraction while prosecuting Indigenous peoples for asserting inherent rights.

Brazilian Indigenous and traditional communities face systematic dispossession as fossil fuel industries destroy livelihoods. Artisanal fisher Bruno Alves de Vega states: "We just want to fish in Guanabara Bay, it's our right. And they are taking our rights." Extractive industries treat territories as sacrifice zones, making traditional practices impossible through contamination and ecosystem collapse, constituting economic and cultural pillage through corporate intimidation.

In Senegal's Saloum Delta, communities report zero meaningful consultation about Woodside's Sangomar fossil fuel project. Residents told Amnesty International they lack accessible information on environmental and socio-economic impacts, don't understand project scope or expansion plans, and explicitly have not consented. This violation of international human rights law and Indigenous rights frameworks continues with impunity.

Every environmental defender interviewed by Amnesty International faces severe risks from fossil fuel disputes. States weaponize legal systems through "lawfare"—criminal proceedings to silence defenders. Indigenous Peoples face arrest, imprisonment, and death for protecting ancestral lands. Amnesty's Candy Ofime states: "States must stop criminalizing environmental human rights defenders." Affected communities condemn the power imbalance and lack of remedy when corporations violate rights.

Fossil fuel projects create pollution "sacrifice zones" on Indigenous lands, whose protectors face consequences while distant decision-makers overrule them. Extraction's entire lifecycle, from exploration through decommissioning, destroys ecosystems Indigenous Peoples protected for generations. Proximity to infrastructure scientifically elevates cancer, cardiovascular disease, and reproductive harm risks. Indigenous communities suffer disproportionate health crises while industry destroys biodiversity areas and carbon sinks that Indigenous stewardship maintained as climate defences.

Despite climate commitments and UN calls to phase out fossil fuels, they still comprise 80 percent of global energy while industry influences policy to prevent change. Indigenous Peoples—least responsible for climate change and protectors of critical ecosystems—suffer the worst consequences while being criminalized for defending lands that could save humanity. Amnesty's Callamard condemns industry greed disguised as development, violating rights with impunity.

Released for Brazil's COP30 vision of Indigenous participation, this report exposes fossil fuels' devastating impact on Indigenous Peoples mounting fierce resistance. Amnesty International demands: a Fossil Fuel Non-Proliferation Treaty, rapid phase-out, just transition respecting human rights, ending defender criminalization, and meaningful Indigenous consultation. Callamard warns: "The age of fossil fuel must end now. It is a major source of human rights violations." Indigenous resistance is humanity's fight for collective survival.

World leaders, human rights activists, climate experts and others have been gathering in the Brazilian city of Belem for COP30, the UN's 2025 climate conference. While countries have pledged to tackle the climate crisis, observers have questioned how they plan to meet their commitments to reduce greenhouse gas emissions while fossil fuel projects continue to expand. Indigenous activists also stormed the summit this week to demand that their voices be heard in the discussions.

The backward thinking of our current government did not go unnoticed at COP30 held in Brazil, November 2025, where international civil society denounced Canada as Fossil of the Day—the first time since 2014 that Canada has been singled out for this dishonour. Fossil of the Day is awarded daily during COPs to the countries "who are the best at being the worst and doing the most to do the least."[75] In a statement, Climate Action Network International said: "Canada receives the award because the new government of Prime Minister Mark Carney has flushed years of climate policies down the drain and is completely 'Missing In Action' at a COP where multilateralism needs to be saved. In addition to the backsliding on policies tackling Canada's climate-destroying pollution, his environment minister Julie Dabrusin has chosen inaction and silence where leadership was urgently needed."

YELLOW QUILL FIRST NATION PRESENTED TO UNITED NATIONS

Yellow Quill First Nation and the Safe Drinking Water Foundation were invited by the Indigenous Environmental Network (IEN) in USA to present the Yellow Quill Story at the United Nations headquarters in New York on April 19, 2005.

The Indigenous Environmental Network looked around the world for some positive news in aboriginal communities and drinking water. They ended up choosing Yellow Quill to represent a community's successful struggle for safe drinking water, even if it did take nine years.

For Yellow Quill's water keeper, Roberta Neapetung, to present at the United Nations and to meet people from around the world was very exciting. The IEN wanted to showcase drinking water innovations suitable for implementation elsewhere as well as collaboration between scientists, engineers and government.

Chapter 28

Alberta's Plan to Poison Sacred Waters

While science is clearly educating and informing people around the world, here at home in Canada, the province with the weakest possible position on environment and Indigenous issues continues what I believe is genocide.

In October 2025, in Fort McMurray, First Nations leaders stood alongside approximately one hundred supporters in powerful opposition to the Alberta government's scheme to "Treat and Release," which is a nice name to dump treated oil sands tailings directly into the Athabasca River system.[76]

For Indigenous communities, this river represents far more than a water source; it is a sacred lifeline that has already suffered through decades of relentless environmental destruction and devastating health consequences from industrial contamination.

Organized by Mikisew Cree Nation and drawing Treaty 8 Grand Chief Trevor Mercredi along with other Treaty leaders, the rally marks the opening chapter of what Indigenous leadership has vowed will be a sustained, uncompromising battle. Their strategy includes aggressive legal challenges and relentless public pressure against what authorities euphemistically call the "treat and release" approach—essentially a plan to discharge toxic industrial waste from oil extraction into waters that Indigenous peoples depend upon for survival.

For years, Indigenous communities living downstream from massive oil sands operations have endured devastating epidemics of rare cancers. These communities hold an unwavering conviction, grounded in their lived experience, that the contamination ravaging their health is directly caused by industrial activity poisoning the water upstream.

Chief Billy-Joe Tuccaro of Mikisew Cree First Nation in Fort Chipewyan delivered an unflinching message to government officials and oil industry representatives gathered at the rally. "People choose their wealth over our health," he declared, exposing the brutal calculation at the heart of the crisis.

His challenge grew even more pointed when addressing the question of Indigenous consent: "I made mention to Prime Minister Carney—if they want our consent, they can make a pipeline from here, take it to Calgary and Ottawa, and they can drink the water first because we're tired of being their Guinea pigs."

This demand cuts to the core of the injustice: if government and industry truly believe their assurances about treated tailings being safe, then decision-makers in Calgary and Ottawa should consume that water themselves—in their own homes, for their own families.

Mikisew Cree elder Margo Vermillion brought deeply personal testimony to the rally, speaking through the pain of witnessing her community's cancer epidemic while contemplating the poisoned inheritance being forced upon those not yet born.

"I was just visiting people here in Fort McMurray that have cancer in our communities. It's our grandchildren that I worry about so much. Our children, yet to be born. What are we leaving them? What are they leaving them?" she said, her voice carrying the weight of collective grief and ancestral responsibility.

Her haunting questions distinguish between Indigenous peoples' unwilling role as victims and government and industry's active role as perpetrators. The community is not choosing this toxic legacy. It is being imposed upon them by those who will never face the consequences.

"If they want our consent... they can drink the water first because we're tired of being their Guinea pigs." This statement from Chief Tuccaro encapsulates generations of colonial violence disguised as resource development. Indigenous communities near the oil sands have been treated as expendable test subjects in an ongoing experiment: how much contamination can human populations withstand before authorities acknowledge the harm?

The answer from Fort McMurray and Fort Chipewyan is clear: No more. If the tailings are truly safe enough to release, then those making the decision should face the same risks. Let government officials and oil executives drink that water every day. Let their children inherit those health consequences.

The rally signals that First Nations will deploy every available tool (legal action, public mobilization, and unwavering resistance) to protect their waters, their health, and their future generations from becoming yet another sacrifice to oil industry profits.

The question is no longer whether the fossil fuel era will end, but whether it will end in time—and whether we will stand with Indigenous peoples or continue to sacrifice them for corporate greed.

Chapter 29

Surge in Reported Water-Related Violence

New analysis from the Pacific Institute in Oakland, California, November 2025, shows a sharp increase in incidents where water is a trigger, a weapon or a casualty of violence.[77]

Violence over water resources reached record levels, continuing a trajectory of steep growth in such incidents over the past quarter century, and especially in the last several years, according to a new assessment from the Pacific Institute. A total of 420 events were reported in 2024, nearly a 20 percent increase over 2023, and a 78 percent increase over 2022. Only 24 such incidents were reported globally in the year 2000.

The findings are drawn from the Water Conflict Chronology, the world's most comprehensive open-source database on water-related violence, created and maintained by the Pacific Institute. The database includes events where water has served as a casualty, a trigger or a weapon of conflict. The data shows a steep upward trend across several categories. In 2024, 61 percent of incidents involved attacks on water infrastructure, 34 percent stemmed from disputes over water access or control, and 5 percent involved the deliberate use of water as a weapon of war.

"The growing number of violent incidents involving freshwater resources underscores the urgent need for international attention," said Dr. Peter Gleick, Senior Fellow and co-founder of the Pacific Institute. "Ensuring access to safe, affordable water for all and safeguarding civilian water systems in accordance with international law are critical to preventing further expansion of violence."

The update shows a rise in cyber-attacks targeting water utilities. Recent surveys of water agencies in the United States and United Kingdom revealed hundreds of attempted breaches targeting systems that manage operations of drinking water and wastewater infrastructure and equipment. It also adds data on violence against environmental and community activists defending freshwater resources, especially in Latin America.

"Our data show that water systems, freshwater resources, and those working to manage or protect them are increasingly affected by violence," said Morgan Shimabuku, senior research specialist at the Pacific Institute. "Continued attention is needed to ensure that economic development proceeds in ways that sustain water resources and the communities that rely on them."

NEVER CUT THE REGULATORS

Founding Windrush Against Sewage Pollution (WASP) in the UK proved strategic. Using retired data scientist Peter Hammond's skills, the group demonstrated water companies were illegally dumping sewage by analyzing industry data, rainfall records, and observable pollution.

The Environment Agency, facing budget cuts to sewage monitoring, couldn't ignore WASP's evidence and resulting media coverage, launching a major investigation into illegal dumping. Campaigner Ned Wells noted everyone now wanted to know about their river's condition. Wells helped secure bathing water status for a Thames swimming site at Port Meadow near Oxford, becoming the country's second river bathing site.

This designation obligated regular bacterial testing during bathing season. Growing numbers of communities recognized this mechanism provided valuable river health information, enabling them to pressure polluters and the Environment Agency using indisputable hard evidence.

Chapter 30

Bill Who?

We have a new threat to Canada's drinking water; its name is Bill! Like all omnibus bills it is undemocratic. Yes, I am referring to Bill 5, Bill 15, Bill 54, and Bill C5.[78]

Bill 5 scraps the Endangered Species Act, dismantles regulations that protect ecosystems and threatens Indigenous rights, while pretenders claim it is *Protecting Ontario by Unleashing the Economy.*

Doug Ford's Bill 5, passed in summer 2025, is the ongoing assault through massive omnibus bills to dismantle decades of environmental protections. Posed as a response to Trump tariffs and the need for economic growth, this bill is actually about cutting corners and silencing the voices of those who stand to protect the lands and waters we rely on. This bill is about giving his government sweeping powers to push through projects without proper oversight.

Bill 5 allows Premier Doug Ford to bypass the Greenbelt Act and make it easier for destructive developments to move ahead without full environmental assessments or meaningful public input. It weakens Indigenous consultation, threatens our water sources, and jeopardizes local farmland and wildlife. We can and should grow our economy responsibly while ensuring the systems we rely on remain intact.

Similarly, Bill 54, also known as the Election Statutes Amendment

Act, 2025, was passed in Alberta and received Royal Assent on May 15, 2025. It is now law and claims that it will improve the consistency and fairness of democratic processes in the province. Bill 54 amends several pieces of legislation, including the Local Authorities Election Act, Recall Act, Election Act, Alberta Senate Election Act, Referendum Act, Citizen Initiative Act, and the Election Finances and Contributions Disclosure Act. While many of these changes may not directly affect municipalities, Alberta municipalities have identified and highlighted those that could. Alberta municipalities are particularly concerned that some of the proposed amendments may result in predetermined outcomes regarding municipal engagement in recall processes and local elections.[79]

Don't think this is only happening in Ontario or Alberta. The British Columbia Assembly of First Nations (BCAFN), First Nations Summit and Union of British Columbia Indian Chiefs issued a joint call for the Government of British Columbia to withdraw Bill 15, the Infrastructure Projects Act, that lets the cabinet craft new environmental assessment processes, automatically issue permits and speed up construction on everything from mines to mills, power plants, schools, hospitals and more.[80]

"Bill 15 is over-reaching and enables the province to bypass permitting processes and expedite environmental assessment for any project they deem a priority," BCAFN Regional Chief Terry Teegee said in a statement.

"The province rammed through Bill 15 without any First Nations input who stand to be adversely impacted. This law may breach constitutional consultation requirements and is not consistent with the United Nations Declaration on the Rights of Indigenous Peoples (UNDRIP)."

The British Columbia NDP (BC NDP) pushed ahead with Bill 15—a dangerous law that gives the premier and cabinet the power to force through major projects without public input, Indigenous consent or environmental safeguards. They say it's about schools and hospitals, but the truth is that Bill 15 gives the BC NDP power to greenlight any project they label "provincially significant"—even if it violates local laws, threatens sensitive ecosystems or overrides Indigenous rights. This isn't just bad policy, it's a direct attack on democracy, environmental protections, and the progress we've fought for in British Columbia.[81]

In Ontario, it's Bill 5, a thinly veiled attempt by conservative Premier Doug Ford to sideline First Nations and open up "special economic zones" where provincial laws would not apply.[82] In British Columbia, it's Bill 15, which will allow cabinet ministers to fast-track projects selected in closed-door meetings, based on criteria the BC NDP keep secret while ramming the law through. Premier David Eby forced every NDP MLA to support Bill 15, including the Speaker of the Legislature, who cast the tie-breaking vote. Ex-NDP cabinet minister Melanie Mark called it "astounding and disheartening" that her former colleagues would "turn their back" on First Nations communities across the province who opposed the bill. I wonder what happened to MPs of all backgrounds representing the views of their constituents.

Eby's former employer, the British Columbia Civil Liberties Association, called it a triple attack on Indigenous rights, environmental protection, and democracy in British Columbia. Even the British Columbia Chamber of Commerce said the rushed, undemocratic process and inevitable legal challenges of Bill 15 actually create more uncertainty for investors, not less.

Why did Eby do it? One clue comes from the only First Nations council in the province that came out in favour of the bill. The Nisga'a treaty government said it supports Bill 15 as part of its push to build the Prince Rupert Gas Transmission pipeline alongside Wall Street investors closely tied to Donald Trump.

"We have been advocating for a more efficient and effective regulatory regime for some time now," read a letter to Premier Eby signed by Nisga'a president Eva Clayton.

This was never about schools or hospitals. Bill 15 is a signal to billionaires in New York and around the world that British Columbia and its resources are here for the taking. Eby unveiled the legislation saying the changes would help him clear bureaucratic delays and permitting holdups by allowing cabinet to designate certain projects as provincially significant and then override environmental assessments and local municipalities to get them built.

An email from Kai Nagata with Dogwood, on June 18, 2025 confirmed my gut feelings:

"The American-owned Ksi Lisims LNG proposal is in a tricky position. Although it has support from the Nisga'a treaty government, it's facing a lawsuit by the Gitanyow Nation, and opposition from Gitga'at, Lax Kw'alaams and Metlkatla. Last week Premier David Eby went out of his way to promote the MAGA-backed gas terminal, ignoring these basic facts:

1. "Ksi Lisims" is a wholly owned subsidiary of Western LNG, based in Texas
2. The profits would go to USA. billionaires, including a major donor, Donald Trump
3. The terminal would be built in Korea by Korean workers, using Korean steel
4. The pipeline would be built by a USA. military contractor, Bechtel

On August 10, 2025, Dogwood further reported that David Eby and the BC NDP were about to approve the Ksi Lisims project, backed by Jeffrey Epstein's pal, Leon Black. If they do, they will be handing the reins to an American billionaire who funded Epstein's child sex abuse ring.[83]

Pipelines are a critical cause of polluted source waters, and polluted source waters are the single most important challenge and financial cost in the treatment process to provide truly safe drinking water.

Federally, Prime Minister Mark Carney's government fast-tracked Bill C-15 during his first one hundred days, which weakens environmental and safety protections for industrial projects. Data shows natural resource lobbyists—oil, mining, and steel executives—had repeated access to Carney and cabinet, while environmental groups were largely shut out of consultations.

Environmental advocates argue this imbalance prioritized fossil fuel industry interests over climate concerns as the rest of the world electrifies and moves towards a cleaner economy.[84] Too many Canadians still do not have access to safe drinking water while our source waters are becoming increasingly polluted. Fires are raging and climate change is real. We need responsible, accountable governance now more than ever.

It did show that when the government has the political will, or maybe incentives to act, they are capable of moving quickly. What Prime Minister Carney should have acted quickly on was fulfilling his governments obligations. As recently confirmed by a federal judge who ruled in favor of three First Nations (two in Manitoba, one in Ontario) in lawsuits against the Canadian government regarding safe housing and clean drinking water. Justice Paul Favel determined the federal government has a duty to provide Shamattawa First Nation and other participating First Nations with safe drinking water. Shamattawa's class-action lawsuit, certified in 2023, represents all First Nations communities nationwide that had drinking water advisories on or after June 20, 2020.[85]

Elizabeth May shared details of Bill C-15, the Budget Implementation Act, an omnibus bill ("omnibus" meaning many different legislative changes stuffed into one bill) explaining how debate and study have been limited. The bill runs to over 600 pages. Much has gone unnoticed.

Elizabeth May was the first MP to point out that the bill gives ministers sweeping powers to exempt people and corporations, at the minister's discretion, from obeying Canadian law. Only the Criminal Code is off the table. Other laws can be broken with the scantest of criteria. Liberal MPs asked where they could find this section of the bill. They were shocked, but they will, as ever, vote as they are told without feeling any responsibility to read the legislation.

Prime Minister Mark Carney and Alberta Premier Danielle Smith signed a memorandum of understanding (MOU) claiming to strengthen federal-provincial collaboration toward achieving net-zero emissions by 2050 while unlocking Alberta's energy resources and creating high-paying jobs. The agreement promotes industrial carbon pricing, private sector clean technology investments, and expanded energy development.

I suggest this much-hyped agreement does nothing toward reaching 2050 net zero emissions and actually represents a profound betrayal of Canada's climate commitments. The MOU allows the federal government to commit billions of dollars to increase oil and gas production in Alberta, build nuclear reactors, construct new pipelines across British Columbia, and lift the oil tanker ban on British Columbia's northwest coast. This abandons Canada's climate policy, violates international and domestic law, and tramples Indigenous rights.

The Honourable Stephen Guilbeault, a passionate and educated environmentalist, resigned from Cabinet in response, demonstrating the personal integrity and political courage previously shown by Jody Wilson-Raybould when she quit Trudeau's cabinet. I now await other Liberal MPs who recognize that the promises they campaigned on now lie in ashes.

When Canada's priority should be demanding urgent climate action, Canada is instead doubling down on fossil fuel expansion, sacrificing our future for short-term political expediency.

MAUDE BARLOW

Maude Barlow, a Food & Water Watch board member and international water rights activist, recounted her fight for water rights. In 2008, UN General Assembly President Father Miguel d'Escoto Brockmann recruited her to establish water and sanitation as human rights. Despite opposition from Canada, the US, Great Britain, the World Bank, and water companies, they brought activists to share stories at the UN.

On July 28, 2010, Bolivian ambassador Pablo Sonam presented a WHO report stating a child died from waterborne disease every three and a half seconds in the Global South. He counted on his fingers, creating a powerful, silent moment in the General Assembly that proved pivotal to their campaign.

Chapter 31

Lobbyists

TurboTax is one of the more well-known providers of tax filing software that Canadians use to file their income tax returns.[86] They, along with a dozen other members, are all served by Tax-Filer Empowerment Canada (TFEC).[87] The organization claims they exist because, though they "can't pay your taxes for you," they're "here to help make the process a whole lot easier and efficient."

I disagree. I believe TFEC exists primarily for the financial benefit of software companies and their shareholders to make big money off the backs of working Canadians.

Why would the government implement this third-party billing system, which was—and still is, in my mind—totally unnecessary?

Andrew Treusch defended the Canada Revenue Agency (CRA)'s handling of the offshore tax scandal of the accounting company KPMG, which involved a controversial tax-avoidance scheme in the Isle of Man. Despite allegations that the scheme was a "sham" designed to deceive authorities, the CRA allowed the case to go dormant after obtaining client names from KPMG. It was later revealed that the CRA made a secret "no penalties" amnesty offer to KPMG's clients, requiring them to pay back taxes and modest interest, which led to scrutiny of Treusch and his officials.

This document shows who was the lobbyist for TFEC, none other than Andrew Treusch, the same man who went to great lengths to protect wealthy investors:

Client name: Tax-Filer Empowerment Canada
Lobbyist name: Andrew Treusch, Consultant
Initial registration start date: 2024-03-25
Registration status: Inactive
Registration Number: 957715-377658[88]

Treusch arranged for meetings on behalf of TFEC with public office holders. So, who is lobbying on behalf of Canadians? Who is standing up to say we don't need this interference or additional expense? In my opinion, lobbyists exist for one reason only: to feed insatiable greed.

The Lobbying Act was revised in 2008, claiming that lobbying is an important part of our democratic system and a legitimate activity. A key duty of government officials is to interact with citizens. The Lobbying Act ensures that Canadians and public office holders know who is communicating with the government and on what subject. I cannot see how this brings MPs closer to Canadians. In fact, I believe it does the opposite. Even if it gave MPs a closer relationship with those who elected them, what is the point when MPs are told how to vote on every bill by their respective party leaders, instead of voting in accordance with their constituents' wishes.

I wonder how anyone had the expertise and qualifications to jump (under Stephen Harper's government) from being appointed Associate Deputy Minister of Environment Canada in June 2008 to Associate Deputy Minister of Public Works and Government Services Canada in September 2009. Then, in December 2012, Treusch was appointed Commissioner of the CRA.

Surely expertise in the area for which they are appointed is necessary. I don't go to an environmentalist or a mechanic for help with my taxes. Neither would I hire a protester carrying a placard for tax advice. I question how a person with such strong ties to lobbying becomes head of the CRA?[89] How can any citizen-led organization, be it a charity or a group filing a petition on Change.org, compete with corporate or political influences through lobbyists?

I believe that at a minimum we are entitled to transparency of who is paying who to lobby, and on whose behalf. But we are also entitled to equal means to compete with such lobbyists and have our own opinions heard. My opinion is that we don't need lobbyists; they do not improve quality of life for Canadians, rather the opposite. You might say this isn't relevant to water. I include it to show readers the power of money. How it is used to increase disparity and to influence government. I believe it is the same increasing group of wealthy individuals and corporations, who put profits over people, who are polluting our waters and ignoring the rights and health of Canadians.

NDP ethics and democracy critic Meara Conway stated that her party is examining various political donations from Nelson Mullins Riley and Scarborough, an American law and lobby firm, to Republican Political Action Committees (PACs) and candidates. This includes contributions to JD Vance's campaign; Marco Rubio; the National Republican Senatorial Committee; and the Fund for America's Future PAC, which supports conservative policies aligned with Trump and his allies.

Conway noted Nelson Mullins has been contracted by the Saskatchewan Party for fourteen years and been paid millions of dollars—$8 million up to 2022 and renewed for $1 million or more over the years since.[90] "We have major fears that public dollars that belong to the people of Saskatchewan have directly gone to firms that have then turned and donated that money to prop up Trump's election machine, and those of his closest allies," said Conway in John Cairns' article.

Why would the Saskatchewan party take money from Saskatchewan people and give it to a firm that props up a Donald Trump MAGA movement hellbent on dismantling our economy? Our public dollars should not be funding politicians who want to make us the 51st state.

Former Prime Minister Justin Trudeau criticized Conservative leader Pierre Poilievre, accusing him of feigning concern over high grocery prices while his top adviser, Jenni Byrne, leads a firm lobbying for Loblaw Companies Limited. The Ontario lobbyist registry lists six employees of Jenni Byrne + Associates as registered lobbyists for Loblaws, though Byrne herself is not listed.

Industry Minister François-Philippe Champagne criticized large grocers for their lack of transparency regarding food inflation and urged-the Competition Bureau to utilize its enhanced powers to address grocery costs.[91] Additionally, the document highlights the significant influence of corporate lobbyists, particularly from the oil industry, on Canadian politics, noting that Big Oil has lobbied the federal government over a thousand times in 2024 alone to protect its profits against climate-related regulations. I invite you to check further to see who is lobbying who.

EDUCATING POLITICIANS

Ash Smith, founder of Windrush Against Sewage Pollution (WASP), noted the water crisis had long been disconnected from public knowledge, but signs indicated a movement was emerging to restore rivers. The issue became so important that Liberal Democrats centered their May election campaigns on river restoration, scoring major gains from Conservatives.

Few people were aware of pollutant cocktails flowing into rivers—the Environment Agency's 2020 assessment showed every English river failed legal health standards. Surprisingly, most dumping was legal, as water firms received permits to discharge untreated sewage into rivers and seas, provided they treated stipulated waste amounts.

Public awareness was finally growing about this environmental crisis.

Chapter 32

Education

For almost thirty years, SDWF has matched sponsors with educators across Canada, delivering over 20,000 Operation Water Drop kits to elementary and high school students. We recognized the need to educate the next generation many years back. Similarly, other charities have education programs, mostly teaching about source water protection. I believe Canadians recognize the need for safe drinking water and source water protection. The people we need to educate are the politicians.

Imagine my horror upon receiving information about the latest government program for drinking water. Another $1.5 million being thrown to the wind.

The honourable Julie Dabrusin, the Minister of ECCC, announced a $635,296 investment in an eco-literacy initiative to be led by AquaAction from the Climate Action and Awareness Fund (CAAF).[92] Their website clearly states their objective: "Accelerate your growth and access our powerful entrepreneurial network with one of our three programs: AquaHacking Challenge, AquaEntrepreneur, and AquaNation."[93] They appear to be all about developing business opportunities, more satellites, more monitoring, even creating washing machine filters. But nothing about water quality, improving water treatment, or source water protection. Simply put, it's all about making money.[94]

Also, the de Gaspé Beaubien Foundation announced they will match the federal investment with a contribution of the same amount, doubling the overall commitment to $1.27 million over five years. They state: "We must overcome the illusion of abundance in Canada and engage young Canadians around improving how we manage our freshwater resources. This landmark funding, paired with generous support from the de Gaspé Beaubien Foundation, lays the groundwork for a transformative, nationwide water literacy campaign. We will partner with leading educational institutions to co-create dynamic, age-tailored learning modules for students from kindergarten through Grade 12. Courses will encourage entrepreneurial thinking and developing job-ready skills to help young Canadians participate in a sustainable blue economy."[95]

Water knowledge is powerful. It sparks curiosity, replaces fear with understanding, and inspires action. The ripple effects of expanding young minds will be felt far beyond the classroom, in everyday choices, future careers, and stronger communities for all Canadians."

This project is one of many created when the Government of Canada invested over $3.3 million in nine grassroots, community projects to curb greenhouse gas emissions and fight climate change. The CAAF was created with an investment from the Environmental Damages Fund, *largely as a result of the highest valued fine* ever received by the program.[96]

The CAAF is designed to support projects that can create middle class jobs for Canadians who work in science and technology, academia, and at the grassroots community level. Since 2020, the CAAF has been funding projects aimed at supporting youth climate awareness and community-based climate action, climate research at Canadian think tanks and in academia, and advancing climate change science and technology.

In the above passages, I have italicized the two sentences which I

155

feel are contradictory and inappropriate. *We must overcome the illusion of abundance*—abundance of what, I ask? An abundance of political denial about the state of water in Canada? An abundance of cover-up about the state of Canada's drinking water or source waters? An abundance of money being tossed to the wind? An abundance of turning water into profit? An abundance of corporations polluting our waters? An abundance of failing democratic governments?

The fund was established in large part by *the highest valued fine ever received* by the program. Who could have paid such a fine I wonder? One of the Big Oil companies? Of course not, they only get hand slapped with miniscule fines.[97] The fund will support community projects that reduce greenhouse gas emissions, with most of the money coming from a penalty Volkswagen paid for violating Canada's diesel emissions standards related to the "dieselgate" scandal. It's part of Canada's CAAF, which aims to fund climate science research, engage youth, and empower local communities to take climate action. Essentially, Canada is turning VW's pollution violation penalty into investments for community level climate solutions.[98]

My next concern is around the duplication of existing programs. I would like to see Julie Dabrusin, the Minister of Environment and Climate Change, enroll for elementary school water lessons from SDWF. I learned she ran for office primarily on concerns about income inequality and government neglect of Canada's urban areas. I wonder what she knows about disparity or inequity based on water quality or government neglect of its responsibilities to provide all Canadians with safe drinking water, or the need to protect its source waters.

Inequality is not limited to urban areas, rather the contrary. Her role should be to question government spending, to question why we don't have safe drinking water for all Canadians. Her role is not to distribute more of our taxpayer dollars to beneficiaries' bank accounts. She could have used the money to fund existing accredited programs into schools across the country. First, I suggest she has a great deal to learn from the existing SDWF as well as other charities and education programs, then all other MPs should follow her.

You can guess what is going through my mind, as I attempted to search for donors to her political campaign. The de Gaspé Beaubien Foundation's specific donations are not publicly accessible through a simple search. I will send her and Prime Minister Carney a signed copy of this book. Of course, I can't ensure they will read it, or act on it. But I have done my best.

MAKING A DIFFERENCE

The Safe Drinking Water Foundation achieved significant impact through its educational outreach program. Executive Director Nicole Hancock discovered at the 2017 Canada Wide Science Fair that their water testing kits had enabled a student from Little Saskatchewan First Nation to qualify for the national competition.

Over 3,000 Canadian schools have utilized their kits, reaching approximately half a million students. Despite funding challenges and limited resources typical of charity work, the organization successfully provided sponsored testing kits to remote First Nations communities.

Their educational tools empowered students to investigate local water quality issues, creating ripples of awareness and action across Canadian communities addressing drinking water concerns.

Chapter 33

Charities

I do not paint all charities with one brush. There are churches, art galleries, dance companies, and religious charities, to name a few. They do not all help the disadvantaged in our society. A charity is defined as an organization set up to provide help and raise money for people or causes in need. Or the voluntary giving of help, typically in the form of money, to those in need. I believe that the charities whose focus is to support those in need only exist because the government is not doing its job.

Legitimate charities are struggling, desperate to source sufficient funds to deliver their much-needed programs. If you feel you are receiving too many requests for donations from charities, it is simply because their donations are falling and demand for their services are growing.

Charities shake their fists and coordinate petitions, insisting that the government must step forward to resolve these issues, all the time knowing they will not. We need to hold the government accountable and demand it lives up to its responsibilities with our tax dollars, by providing the programs and safe drinking water to support all Canadians.

People also believe that charities are organizations that should operate in a "less is more" fashion. The less staff are paid, and the lower the overhead, the better any charity is perceived. Unfortunately, the charitable sector does little to push back against this toxic narrative. One does not need to think too hard to figure out why: they don't have the staff and resources to deliver their programs, never mind pushing back. There is no place for charities to spend on luxurious offices etc.; their focus and mission should always be reflected by how they spend their donations. Numerous foundations, who typically fund charities, are closing their doors. There is less money to go around, and there is a greater need for their services than ever before.

I cannot understand or justify that corporations can lobby politicians and gift huge donations to political parties, buying their influence. Wealthy individuals, corporations and media are all empowered and accepted forms of persuasion and coercion to influence government toward their own agendas. Yet charities must refrain from partisan activities. Charities must resort to organizing petitions or protests to influence government decisions. Charities must not operate any form of profit-making business.

Yet charities are expected to attract staff when they cannot raise the revenues necessary to fulfill their much-needed programs, unless the charity receives funding from the government. I see two levels of charities.

I submitted a FOI request to the federal government in July 2024 to see how many charities had evolved to deal with drinking water issues in Canada since Dr. H_2O and I founded SDWF in 1996. Statistics were available from 2003.

Canada had 80,581 registered charities in 2003, with sixty being related to water—or at least containing the word *water* in their name. In 2024, overall charities grew to 85,519 and eighty of those had water in their name. Of twenty new charities, only three are focused on water, the other seventeen are dance groups or churches or arts with water in their name, e.g. Church of Living Waters. I find it discouraging that only three new charities appeared during the two decades that are focused on drinking water.

The only charity focused on safe drinking water remains to be

SDWF. KOW focuses on protecting source waters, which often leads to them initiating legal challenges and their strong advocacy.

While "charity" and "nonprofit" are often used interchangeably, a charity is a specific type of nonprofit organization (NPO) focused on charitable purposes, while non-governmental organizations (NGOs) are NPOs that operate independently from government oversight, often with a broader, international scope. Or it can be explained that an NGO is a voluntary group of individuals or organizations, usually not affiliated with any government, that is formed to provide services or to advocate public policy. However, an NGO must also be a registered charity with the CRA in order to issue tax receipts. Charities often donate to NGOs.

While many NGOs are also advocates including Amnesty International, Greenpeace, and the Centre for Reproductive Rights, I have crossed paths with too many who refuse to participate in advocacy, and coincidentally they are funded by government. And then there are the NGOs who simply choose not to advocate for their mission, which I simply do not understand and so I ask why not?

Genuine charities that work in the trenches are often held in squalor by the CRA, who threaten to take away their charitable license for any political *interference*, depriving them of creating business income as they are not supposed to run a business for profit, and ignoring their protests. The government carries on with business as usual despite being informed of the critical issues which they refuse to address. The SDWF has been educating and informing government on all levels and all aspects of what needs to change to provide all Canadians with safe drinking water, for almost thirty years.

I am also aware of charities that receive substantial revenue from government departments. I believe that when any government funds a registered charity or an NGO, they are not funding the cause; they are silencing or controlling the organization, and hoping they look good in the process. It is no different from how large corporations buy up smaller businesses that may impact their agendas.

I also question why the CWN has been established as a national nonprofit, though it is funded by ISC? The benefits of being nonprofit can include tax exemptions on income and property, eligibility for grants and public funding, increased public credibility and donor confidence, and the ability to collaborate with other organizations. Nonprofits also provide a sense of purpose for their members and staff by allowing them to work on meaningful social and community issues. Operating as a nonprofit the board of directors still require liability insurance, protecting the founders and directors from personal responsibility for the organization's debts.

I share with you a report released by the CWN on November 27, 2025, National Infrastructure Assessment Report, Building Foundations for Tomorrow.[99] I quote, "Solid waste infrastructure is generally in good physical condition across the country." I reference sewage over fifty times in this book. I have an entire chapter devoted to the topic. Do you feel sewage is not a problem in this country? If you do, I need to take more lessons in how to write.

I call this report the epitome of waste when reading the two paragraphs on water and wastewater. More money being thrown to the wind. The report did not cover the huge issues that municipalities, towns and provinces are facing such as overload of sewage combined with insufficient or ineffective water treatment facilities, or the cancer-causing asbestos cement pipes delivering cancer from the drinking water taps of homes across the country, all of which must be removed, covered in more depth later.

Charity salaries should not be considered frivolous allocations, but rather savvy investments. Or are they? Projects should make a difference in accordance with the mission of the charity. Charities are governed by a volunteer board of directors, who are responsible for ensuring they have the resources they need to meet their mission. But what about board members who benefit significantly from their positions? For example, board members who receive honorariums for delivering workshops that could or should, in my mind, be delivered by paid employees. Or a board member whose primary goal is to pad their resume. Or boards who sit back and leave all responsibilities to the executive director, coasting in the limelight but rarely attending board meetings. Or the board member who benefits from billing a charity for services, items or gifts to the staff of the charity? Or a board that stagnates, maybe for decades, without attracting new board members. Or the board members who benefit from fine dining or celebrations, along with the staff. I find these to be corrupt situations. There are also board members who sit back for the ride, sometimes for decades, never advocating. Silence is a weapon and it benefits the oppressors.

While the great majority of board members volunteer and donate generously to their organization, the door is open for those who take advantage of their position. Just like the Koebel brothers who were trusted with carrying out accurate and honest water testing at Walkerton, some people have no conscience, or perhaps no training, and that can include board members.

I have arrived at the conclusion that very few charities achieve what they set out to do. Too often they become futile shadows of their initial goal, often wasteful, self-serving, paying employees too much to achieve too little, rarely advocating for their cause. Just like the corporations I criticize for putting profits over people, many charities pay themselves before paying the cause. Too many charities duplicate efforts and thereby costs with similar goals which they could achieve with more success by working together. Corruption exists in charities as it does everywhere else in society, and it exists with the full consent of the government. I reiterate, charities only exist, as do unions, because government fails to act in the best interests of its people.

How does a person report a charity? I reported a charity's misdemeanours to the CRA National Intake office. When I received no response, I sent a similar letter to the National Leads Centre of the Canada Revenue Agency. Both letters were received, confirmed by registered mail. Yet once again, like FOI requests which vanished into the ether, I received no response.

Water Watchers began as a grassroots organization founded in 2007 to oppose Nestlé Water Canada's local water bottling operations in Wellington County, Ontario. We need all the opposition, all the advocacy to bring awareness to the deplorable government approval of companies such as Nestlé who pay a minuscule amount for the water they then sell for millions of dollars. Today, Water Watchers have become a unique resource to grassroots leaders on the frontlines of water protection

Righting Relations is another strong, much needed charity making significant progress, offering adult education for social change.[100] They are a movement of adult educators and community organizers who are working towards decolonization and radical social change across Turtle Island.

A coalition of NPOs is lobbying for a "more equitable and effective federal funding system," but their demands raise serious questions about accountability and ulterior motives.

The coalition, supported by Imagine Canada and funded by Canadian Women's Foundation and Cooperation Canada, claims "outdated funding practices limit their effectiveness." Their solution? Dramatically reduce government oversight and reporting requirements.

Their specific demands include replacing contributions with grants to eliminate stricter accountability measures, allowing unrestricted movement of funds between budget lines so money allocated for programs could be redirected to salaries without approval, longer funding terms with reduced approval wait times to lock in multi-year commitments with less scrutiny, and a streamlined system that reduces administrative work across multiple federal funders.

While framed as "reducing administrative burden," these changes would decrease transparency for how taxpayer dollars are spent. Should organizations receive ten-year funding commitments with minimal oversight? Should they freely redirect program funding to administrative costs without justification?

One must ask: who is truly pushing for these changes? The nonprofits receiving funding certainly benefit from reduced reporting and greater financial flexibility. But what about the vulnerable communities these organizations claim to serve? For organizations working with populations historically failed by inadequately supervised programs, shouldn't we demand more rigorous oversight, not less? True partnership doesn't mean eliminating accountability—it means transparent collaboration with robust safeguards protecting both funders and communities served.[101]

Please remember Water Watchers, Righting Relations, and KOW, who are all strong advocates for their causes, like the SDWF, and compare them to my next two examples.

PROTECTING OLD GROWTH FORESTS

Near Cathedral Grove, a major tourist attraction on Vancouver Island, the Cameron Valley Ancient Forest ("Firebreak") was originally preserved as a firebreak and elk/deer habitat. In 2004, BC's liberal government deregulated it, enabling logging of one-third of its 150 hectares by 2012.

The **Ancient Forest Alliance**, stepping into the government's abandoned role as forest protector, has campaigned relentlessly for its preservation. Their advocacy recently secured temporary deferrals from current landowner Mosaic Forest Management, but permanent protection remains critical—99 percent of BC's ancient Douglas firs are already gone.

This demonstrates how environmental organizations must fulfill the conservation duties governments neglect, fighting to save irreplaceable ecosystems from destruction.

Chapter 34

The Saskatchewan First Nations Water Association (SFNWA)

Toward the end of my previous book, *Water Confidential*, I exposed the shortfalls of the government allocating tax dollars under the auspices of First Nations drinking water. I shared my perspective about how ISC support of SFNWA has abused these funds. Its most significant project is organizing an annual gathering, with teepee raising, entertainment, dining, and partying.

Thanks to a FOI request—which I persisted in filing—I discovered that, apparently, ISC had no problem with SFNWA charging attendees to enter the gathering, despite the fact that they received ample funding from ISC to host the event.[102] I call that profiteering. SFNWA is both an NPO and a charity. It is registered as a nonprofit, meaning its goal is to benefit First Nations, not to profit from its activities. It also has a charitable status, which allows it to apply for funding from foundations, etc.

As if charging attendees on top of receiving ISC funding wasn't providing sufficient profits to enable business class flights, stockpiling money and paying inflated salaries, an SFNWA Instagram post reads, "We are so grateful to SaskWater for sponsoring the keynote address at the 2025 SFNWA Conference by the inspiring Cadmus Delorme. His presentation was phenomenal, thought-provoking, and (at times) hilarious."[103]

Having lived in Saskatchewan for nearly three decades, I am not the least surprised at how SaskWater still throws its taxpayer money to the wind. SaskWater's Spudco initiative in the late 1990s was an ambitious attempt to develop a large-scale potato industry in Saskatchewan, including the creation of the Lake Diefenbaker Potato Corporation and construction of storage facilities to support a planned french fry plant. The project ultimately failed, resulting in bankruptcy and costing Saskatchewan residents over $12 million. How can a Crown corporation justify growing spuds or sponsoring keynote speakers when their website statement is "As Saskatchewan's commercial Crown water utility, we partner with communities, First Nations, and industry to provide safe and reliable water and wastewater services."[104]

I noticed SFNWA using AI to create text for its website. SFNWA posts profiles of its board members, and in the spring of 2025, when they profiled board member Ernie Jimmy following his experiences, they forgot to edit out the AI comments. I believe the SFNWA also uses AI to generate the lessons they provide for WTOs to gain their Continuing Education Units (CEUs). I wonder why a charity in Saskatchewan is creating courses for water systems operators to earn CEUs—other charities are probably doing the same in other provinces. Why not one program for CEUs for all Canadian water systems operators? After all, all First Nations communities are under federal jurisdiction, not provincial. I feel the charity is doing the work of the government.

SFNWA seems to change its mind about whether to share its audited financial statements on its website. According to documents shared, SFNWA received funds from ISC in 2022 and in 2023 for a total of $1,351,000. It received funds again in 2024 and a commitment for 2025. I estimate well over $2 million in four years. At the time of writing, the SFNWA is listed on the charity website in 2024, but no financial statements are shown there.[105]

I can find no evidence that SFNWA takes direct action to improve water quality or protect source waters in First Nations communities. Until recently, it had no First Nations people on staff. It appears to focus on offering workshops to help WTOs earn their required CEUs— but what good is that if the treatment plants in the operators' home communities are largely ineffective?

The SFNWA does no advocacy whatsoever. It is, in effect, an extension of ISC, delivering government programs or, perhaps more accurately, government attitudes.

In a package I obtained via FOI, in an email from the executive director of SFNWA, to her ISC funder, she included a wink emoji in her message about them both skipping a government session. This makes me think of *Monty Python* skits: nudge nudge, wink wink! I question her flirting and their coziness.

My latest FOI request to SFNWA yielded a response that identified ISC had hosted an event titled, "Proposed Updates to the Protocol for Centralized Drinking Water Systems in First Nations Communities: A facilitated discussion with Indigenous Services Canada" to be delivered by the Strategic Water Management employees, Andrea Cherry and Meghan Myles.[106]

The session was twenty pages of intense scientific jargon, with a healthy dose of mathematics, and lots of government speak. Most WTOs whom I have met have struggled with the math required to update their CEUs. My head hurts reading this. I can only imagine how hard it is for WTOs. Yet ISC asks for their feedback on ISC proposed updates. ISC calls it a facilitated discussion on Strategic Water Management. I call it a dictatorial colonial speech that the facilitators can claim was consultation. This project has cost thousands of dollars in civil servant time alone, never mind the cost of hosting such a conference. Yet the operators are paid little above minimum wage and returning home to mostly ineffective water treatment plants.

I learned from the same package of information received from my FOI that the Assembly of First Nations (AFN) initiated development of technical standards for the design and operation of drinking water systems under their authority.

I sent an online inquiry to AFN. They, too, are dependent on government funding. I reached out, hoping to obtain more information on how AFN is moving toward assuming responsibility for their people's drinking water under Bill C-61. The AFN response:

> "Thank you for your email and questions. The Assembly of First Nations (AFN) is a national advocacy organization that receives mandates from the First Nations in Assembly. Our role is to advance these mandates including new water legislation for First Nations. In reviewing some of your questions, the AFN does not develop guidelines rather we support the autonomy of First Nations and Regional First Nation organizations in the development and implementation of their preferred water and wastewater initiatives including development of regulations etc. At this time, we are not able to provide a direct quote. We continue to work with and support our First Nations to have clean and safe drinking water as a basic human right."[107]

Every time I read the SFNWA newsletter I find their use of funds to be highly inappropriate. When so many First Nations communities in Saskatchewan do not have safe drinking water, and SFNWA receives millions of dollars from government agencies, I cannot justify spending that money on the entertaining events they host.

One of their updates on Facebook, read, "SFNWA Board Member, James Cappo from Muscowpetung Saulteaux Nation #80, put together a team of water operators from File Hills Qu'Appelle Tribal Council in TREATY FOUR to compete in the Echo Lake Plywood Regatta. Thank you, James, Solomon, Rodney, Curtis and mascot Bannock for representing the SFNWA in a good way."[108]

A few weeks later, the SFNWA sponsored another event, this one quite different from the first. The Bridge City Warmth Community Dinner and Distribution Event is undoubtedly a much needed and worthwhile cause in Saskatoon, but it falls outside the SFNWA's stated mission: Strengthening water management education, certification, and collaborative networking for First Nations water professionals in treaty territories.

Susan Blacklin

"Bridge City Warmth Community Dinner & Distribution Event, our amazing BCW [Bridge City Warmth] volunteers are on site and working hard to bring together an incredible Dinner & Distribution Event, proudly sponsored by The Saskatchewan First Nations Water Association. Come down and join us for a warm meal, great company, and community spirit in the heart of Saskatoon."[109]

SWITZERLAND'S ELECTORAL SYSTEM ENCOURAGES REFERENDUMS

Switzerland's electoral system enables any citizen to launch optional referendums against parliamentary bills, requiring minimum consensus for laws to survive challenges.

Annual election and referendum costs total CHF 233 million (Swiss francs)—approximately $75 CAD per registered voter or $45 per resident. Canada's 2025 federal election cost $570 million, about $20 per registered elector or $12.50 per resident.

This means Canadians would need approximately $60 more per voter to implement Switzerland's accountability system through direct democracy referendums.

Chapter 35

Water First

I learned of another charity, Water First, located in Ontario. Like SFNWA, Water First seems to have a cozy relationship with ISC. According to its website, Water First claims to be "Canada's leading charity dedicated to working in partnership with Indigenous communities to address local water challenges through education, training, and meaningful collaboration."[110]

I love the passion and desire that Indigenous youth bring to help their people obtain safe drinking water. I know how their brains soak up information, especially that learned via a hands-on process. Dr. H_2O taught hundreds of First Nations water operators. But, like all people, youth are only as smart as what they have been taught. If they are not taught critical thinking and scientific analysis, it requires more than biology 101 to become experts in the field of safe drinking water.

Water First is a registered charity, although it is funded in large part by the government, but also receives funding from corporations, foundations, and individuals, including AFN.

My question is, at what point does a charity, an NPO or an NGO, become an arm of government? How much of their revenue needs to come from the government before it becomes a mouthpiece for the government?

Water First has received a commitment from ISC to fund 50 percent of its revenue, almost $5 million, from 2024 to 2029.[111] Does that make them an extension of government? Could that be why they do not embark on advocacy? Is government funding a form of hush money? I am struggling to see why ISC needs to support this charity. I would like you to consider the following facts:

- ISC funding for Water First supports administrative salaries; it does not fund the training or education of WTOs. [112]
- Water First revenue exceeds their expenses by nearly $1 million, suggesting it did not need additional government support.
- They spend over half of a million dollars on fundraising in a year.
- It also spends nearly $500,000 per year on management and administration. The amount it will receive from the government for administrative salaries in 2025–2026 is almost enough to cover its 2023–2024 administration and management costs—meaning the government will soon be funding nearly all of its overhead.
- Water First spent over $363,000 on travel and vehicle expenses in a single year.
- They currently have over $7.3 million in assets (or around $5 million after you subtract liabilities).
- Professional and consulting fees total nearly $700,000, a significant amount given its already well-compensated staff.
- In 2023–2024, staff and volunteer training cost just $38,362, yet for 2024–2025 the organization received $80,000 in federal funding exclusively for internal staff training!

All of the above points are drawn from Water First's financial statements.[113] Now, let us look at how those numbers translate into the colourful claims featured in their annual reports.

Founded in 2010, Water First has been operating for fourteen years as of October 31, 2024. During that time, the organization has reported 207 certifications earned by interns, averaging approximately fifteen per year. A total of 9,379 students participated in the School Water Program, or about 670 per year. By comparison, the SDWF reached 1,902 students in 2024 alone through its water education programs. The Drinking Water Internship Program has produced fifty-four graduates, or an average of fewer than four per year!

The School Water Program team has delivered 152 water science courses. This appears to be a relatively recent initiative, so I'm not going to average it. In 2024, forty-eight courses were delivered. Students earned 105 high school credits through the summer credit course, including twenty-one in the past year. If 152 courses were offered and 105 credits completed, the completion rate was approximately two-thirds.

In addition, the organization has delivered eighty-three environmental water workshops, or roughly six per year, and conducted thirty-seven environmental water projects, averaging fewer than three annually.

From the Water First audited financial statements, there are many questionable points I feel warrant further investigation, such as:

They have over $1.2 million in a contingency fund, which is about a quarter of their money. (That seems excessive to me.)

While they only bought their building about three years ago (for $829,446), they are constructing a new building starting in May 2025.

There were twenty-one drinking water internship participants in 2024 and these participants earned twenty-six provincially recognized water treatment certifications. My guess is that they didn't all pass, and that probably those who passed got two certificates (water treatment and water distribution), so I am guessing that thirteen got certified and eight failed or dropped out.

I cannot help but compare how Dr. H_2O's dedication and passion resulted in over forty communities benefitting from improved, effective water treatment plants, benefitting thousands of First Nations residents, and how he trained literally hundreds of WTOs in numerous communities. I find it totally unacceptable that Water First has not improved one water treatment plant. I cannot find one community that has benefited and now has drinking water that meets or exceeds the guidelines established by the WHO. I estimate that everyone who Water First has trained is working on an ineffective, outdated system that cannot meet WHO standards in their home communities.

I would expect a list of deliverables, or perhaps measurables, to be attached to their funding, but that is simply not the case according to information obtained via my FOI request.[114] As you see from the required "Outcomes," it is all about expanding and growing Water First, nothing at all about improving drinking water quality or protecting source waters.

As far as I can tell, there are many similarities between Water First and SFNWA. Both organize celebratory gatherings from your tax dollars. This is from an email sent by John Millar, executive director and founder of Water First, to Minda, an employee of ISC, on September 13, 2024:

> "Sorry for the radio silence lately, Minda. We have a staff gathering all next week, together with members of the board and Indigenous Advisory Council—it's nearly fifty people coming together from all over Canada. We've been a bit up to our eyeballs lately."

If you tally up their staff, board members, and Indigenous Advisory Council, they make up around fifty people. Is this who enjoyed the event? I wonder how this event, like the SFNWA annual event, improves drinking water for even one community.

They show their achievements in their annual report as coloured bubbles. I encourage you to visit their website to read their explanation of the work they do on drinking water.

I wanted to find out more about this organization, so I signed up for a webinar they offered on November 21, 2024. The zoom invitation arrived in my in box: "Hi Susan, Water First's virtual event Voices Behind Safe, Clean Water is starting now! Click the button. "

Thirty-five silent staff? No chat open. Q&A controlled.

Teresa (the host) responded to my question, "Can you tell me what you test for in the source waters, and do you test the treated waters?" by telling me that she would contact me after the Zoom, but she never did. I followed up. Still no reply.

When anyone tries to shut you out or avoids answering your questions, I feel they are either incompetent or hiding something. I attempted to talk with another employee, but she could not answer my questions either. Eventually, I emailed Teresa. Then a month later, John Millar, executive director and founder of Water First. I was not surprised that they both chose not to respond.

Canada's clean water challenges have been interpreted as merely an inconvenience by most Canadians; others have educated themselves and learned it was often a life-threatening emergency for affected Indigenous peoples. Canadians want to help. In response to this very dilemma, a coalition of Rotary Clubs from Districts 7080 and 7090 in Ontario embarked on an ambitious collaborative effort, claiming they conducted extensive research and dialogue before pursuing an innovative solution. I question how extensive their research was, or how innovative their action.

Working in partnership with an international counterpart in Buffalo, this group secured a Global Grant worth $115,000. They chose to donate this money to Water First. Maybe they assumed that if ISC supports Water First, then it must be good. I challenge Rotary clubs 7080 and 7090 to show me a community who has improved the quality of drinking water as a result of their donation.

I wonder if John Millar, executive director and founder of Water First, has magnets or horseshoes in his pockets, and uses them to attract celebrities to Water First. He received a special medal at a ceremony at Rideau Hall in Ottawa in February 2025, for "contributions in nonprofit management working alongside Indigenous communities to address drinking water and environmental water challenges through education, training, and meaningful collaboration."

Millar was one of the recipients invited to Ottawa to receive the medal directly from Governor General Mary Simon, who is Canada's first Indigenous Governor General.

"This is all about acknowledging the tremendously hard work of the interns and staff, and really celebrates the model that we strive for," says John Millar. "We're proud to be recognized for the way we work: a partnership of Indigenous and non-Indigenous people coming together to solve complex problems and create better solutions."

I fail to see what complex problems Water First are solving or what better solutions they are developing. Not one community has improved quality of drinking water, and no source waters are protected or less polluted as a result of Water First's work.

I suggest that as Canada's charities appear to be operating without necessary government oversight, more people will make uninformed decisions about where to donate their money.[115]

CBC's *Fifth Estate* aired a documentary declaring charitable activities contrary to Canadian foreign policy run afoul of tax rules.

ISC filed an investigation into the Federation of Sovereign Indigenous Nations following allegations and complaints filed with the department's Assessment and Investigation Services Branch in March 2024.[116] The subsequent forensic audit by KPMG uncovered $34 million in questionable, ineligible, and unsupported spending across a five-year period, raising critical questions about financial accountability within Indigenous governance structures in Saskatchewan.

Could it be the department faced mounting pressure to demonstrate oversight and transparency? I suggest ISC may be opening up a much needed national inquiry: first, into ISC funding of questionable—nudge nudge, wink wink—relationships that do nothing to improve quality of life for Indigenous communities;[117] second, into where the money has gone in many communities, money for which ISC has no records, such as the $30 million to Neskantaga First Nation.

ENVIRONMENTAL ART

Los Angeles-based artist Sean Yoro used his paddleboard to paint semi-submerged murals on shipwrecks, abandoned docks, and sea ice. He created portraits of women and hands emerging from water, fully visible only at low tide before disappearing with changing tides.

Yoro, who grew up surfing in Hawaii, explained environmentalism drove his art, calling it his "kuleana," or responsibility. His viral murals raised awareness about dying coral reefs, climate change, and sustainability.

Using only eco-friendly materials meant his works were impermanent, which he found enhanced their environmental message. Working from his eleven-foot paddleboard while managing currents and tides posed significant challenges, but he remained committed to advocating for environmental issues.

Chapter 36

Charity Intelligence (CI)

Water First proudly displays the CI logo. CI's overall star ratings are based on an assessment of donor reporting, financial transparency, funding need, and a concept called "cents to the cause," as well as demonstrated impact.[118] Five-star rated charities, based on these combined metrics, are listed in CI's one hundred highest rated charities report, which includes Water First.

The demonstrated impact rating looks at only one aspect of charity performance: for every dollar donated, the measurable return to clients and society. For example, when comparing two charities where one saves lives for $100,000 each while a second charity may save lives for $20,000 each, impact donors will support the second charity because it creates five times the impact per dollar.

I am also struggling to see why CI validates Water First as a charity which donors should support. I would like to see CI conduct a financial comparison of Water First to SDWF.

Some call CI a "charity watchdog," yet I wonder why CI doesn't rate Water First according to the number of communities (zero) or the number of people (zero) who have safe drinking water as the result of its programs. I liken CI rating of Water First, compared to their analogy of saving lives, that Water First are only assessing whose lives should be saved. And why isn't advocacy a requirement for the CI rating of a charity such as Water First?

I suspect that CI bases its decisions on AI.[119] I hold high regard for Imagine Canada. They work with, and on behalf of, the nonprofit sector to help build a just and more fair society for all. Through advocacy efforts, sector research, capacity building resources and services Imagine Canada supports charities and nonprofits so they can better serve communities both here and around the world.

Imagine Canada hosted a panel discussion on the ethical considerations that charities and nonprofits need to consider when embedding AI technologies within their organizations. I don't recall CI or Water First participating.[120]

ENVIRONMENTAL DEFENSE FUND SPEAKS OUT FOR WATER

Ontario's Water and Wastewater Public Corporations Act threatens public water despite government assurances. The legislation never mentions public ownership, enabling privatization any time.

Ministers can designate water corporations for municipalities without consultation, while granting immunity from liability to officials—contradicting Safe Drinking Water Act protections established after the deadly Walkerton crisis.

This dangerous gap between political promises and legal reality removes accountability safeguards while opening doors to privatization. Environmental advocates demand repeal of this Act, emphasizing that water is a public right requiring genuine protection, not legislative loopholes that betray communities.

With government failing its duty, organizations must defend water as the essential public resource it must remain.

https://www.instagram.com/reels/DSViTrQkVtt/

Chapter 37

Artificial Intelligence (AI)

I believe that AI has tremendous potential for assisting in solving the issue of ensuring safe drinking water. Surely it could be trained to assess specific analysis results of source water to determine or qualify if a treatment system is effective and then analyze the treated water data to detect if any health threats remain. Maybe scientists and engineers can be some of the first jobs to be replaced by AI. But then, if we analyzed both source and treated waters, scientists could come to the same conclusion without AI. And if we had national drinking water regulations, we could protect the health of Canadians.

Maybe we need it to be recommended by AI for it to happen.

Canada was the first country in the world to introduce a national AI strategy and has invested over $2 billion since 2017.[121] I can see so many great opportunities for AI—but in my opinion it is a dangerous and potentially misleading tool if it is not regulated. I see AI rather like drinking water or the Internet; unless AI is regulated, nothing protects it from being used for exploitation.

Evan Solomon, a seasoned broadcaster and newly elected Toronto Centre MP, has been appointed as Canada's first-ever Minister of Artificial Intelligence. This role, introduced shortly before the 2025 election, suggests Solomon will oversee significant aspects of the economy and national security related to AI.

Mark Carney, Solomon's superior, advocates for extensive AI use to build the "economy of the future," encouraging businesses to adopt AI and develop the necessary infrastructure.[122]

Solomon, as the minister of AI and digital innovation, will likely oversee the establishment of a digital transformation office to enhance government services and suggests that AI could address service backlogs and improve delivery times, providing better services faster. Could AI speed up responses to FOI? I don't think I should hold my breath.

Prime Minister Carney's platform proposes a tax credit to encourage AI adoption by small and medium-sized businesses and aims to connect Canadian researchers and start-ups with businesses nationwide to accelerate adoption.

The Liberal platform also has an emphasis on building Canadian-owned AI infrastructure, including data centres and high-speed and reliable communication networks.[123]

"That's crucial," says Benjamin Bergen, president of the Council of Canadian Innovators, "for Canada's sovereignty." He continues, "Imagine when we were building the railroad, which built this nation from east to west, if we had allowed foreign agents to determine what goods we could put on the railroad, what goods could travel, who could travel."

This all sounds so forward thinking, but what I don't hear is the much-needed word *regulation*. I am confused. We allow foreign investors to build the KSG pipeline? Imagine a train running down tracks with no regulations. The way the internet has had no repercussions against the dark web or all the fake news and outright lies that are pathetically labeled as misinformation. I have often compared the lack of drinking water regulations to driving on a highway with no speed limits, and the same applies to trains or to AI without regulations.[124]

A United Nations International Labour Organization report reveals that jobs traditionally held by women are more susceptible to the impact of AI than those held by men, especially in high-income countries.[125] The report, dated October 28, 2025, indicates that Amazon announced it will cut about 14,000 corporate jobs as the online retail giant ramps up spending on AI while cutting costs elsewhere.[126] After a group of attorneys were caught using AI to cite cases that didn't exist in court documents, a lawyer was told to pay $15,000 for his own fraudulent AI in several briefs. The judge wrote that AI is "much like a chainsaw or other useful [but] potentially dangerous tools, one must understand the tools … and use those tools with caution. … It should go without saying that any use of artificial intelligence must be consistent with counsel's ethical and professional obligations."

In other words, the use of AI must be accompanied by the application of actual common sense and intelligence. One of the biggest problems facing society today is misinformation. Without regulation, AI, the internet, and all social media can alarmingly misinform the masses.[127]

I was amazed to read that a Nobel prize winner, Dr. Geoffrey Hinton, a recent co-recipient of the prestigious Nobel Prize for Physics, announced a generous donation of $350,000 to Water First Education & Training, the sum of half of his prize winnings. Dr. Hinton, a computer scientist and professor emeritus from the University of Toronto, was awarded the prize for his groundbreaking contributions, along with John Hopfield, in the field of AI. I wondered if this was about furthering AI rather than the peer-reviewed sound science of water quality.[128]

I think of two scientists who passionately improved water quality during their lifetimes. The late Dr. Dave Schindler, former chairman of the SDWF, led or contributed to over 300 scientific publications, affected policy change in countless contexts, mentored hundreds of students and colleagues, and received dozens of honorary degrees and international awards, including the Nobel-comparable Stockholm Water Prize. Or David Suzuki, a favourite collaborator for Dr. H_2O, a renowned scientist, broadcaster, and environmental activist. He did not win a Nobel Prize. He did receive the Right Livelihood Award, often called the "Alternative Nobel Prize," in 2009. The two Daves, both incredible, honourable scientists who made great accomplishments in water quality.

But then I read Charlie Angus' memoir, *Dangerous Memories*, where I learned more about Geoffrey Hinton. "One of the Godfathers of AI," he had warned that the capacity of AI to create mass unemployment, supplant human decision-making, undermine political conversation, and launch militarized death machines poses a more urgent threat to humanity than climate change."[129] I loved reading Charlie's memoir, realizing that it's not just me who has dangerous memories. It wasn't Charlie's opinion; it was reported by Reuters in 2023.

"Artificial intelligence could pose a 'more urgent' threat to humanity than climate change," AI pioneer Geoffrey Hinton told Reuters in an interview. Geoffrey Hinton, recently announced he had quit Alphabet (GOOGL.O) after a decade at the firm, saying he wanted to speak out on the risks of the technology without it affecting his former employer."[130]

It appears obvious that being a highly qualified scientist, Dr. Hinton is fully aware of the potential downfalls of AI and the need to conduct thorough research. I am completely baffled as to why he would then endorse an organization like Water First, and the CI who publicly encourage donations to Water First.

ITALY LEADS IN ETHICAL AI GOVERNANCE

Italy has become the first country in the European Union to pass a comprehensive law regulating artificial intelligence, marking a major step ahead of broader EU legislation.

The new law championed by Prime Minister Giorgia Meloni's government aims to make AI transparent, human-centric, and safe, while also promoting innovation, cybersecurity, and privacy protections. It introduces criminal penalties for harmful AI misuse (including deepfakes and fraud), imposes safeguards across key sectors like healthcare, education and justice, and sets limits on AI access for children without parental consent.

This legislation positions Italy as a leader in ethical AI governance in Europe.

Chapter 38

Artificial Intelligence (AI) Needs Water, Too

Benton Harbor, located on the southeast shore of Lake Michigan and renowned for its vacation atmosphere and beautiful beaches, is one of the poorest cities in Michigan. In recent years, the area has struggled to secure funds for essential infrastructure, particularly for its water supply, which was found to have dangerously high levels of lead.

Facing a $2.5 million annual deficit in operating its water provider system, the city has turned to deep-pocketed multinationals to help bring about change. Last year, reports emerged that an unnamed company was seeking to build a $3 billion data centre on 280 acres of land east of the city. Not everyone backs the move, including the Benton Harbor Community Water Council.

A vast plot of land in the municipal district of Greenview, Alberta, is the potential site for "Wonder Valley," what would be the largest AI data centre complex in the world. Celebrity investor Kevin O'Leary, also known as Mr. Wonderful, is a Canadian businessman, television personality, and actor, who is perhaps best known as a panelist on the reality series *Dragons' Den* and *Shark Tank*. Not everyone loves Kevin O'Leary.[131] Jon Stewart Called him an "a**hole" on the Daily Show. This label was applied when it became apparent that Kevin O'Leary is supposedly bankrolling this Alberta Wonder Valley.[132]

The tech hub, which will include buildings that store and process digital information, will come with a total investment over the lifetime of the project of more than $70 billion, according to a news release.

While AI and cloud computing demands continue to grow, technology companies are increasingly interested in the Great Lakes region due to its cooler climate and abundant water resources. I am guessing that Alberta source waters are as cold as, or colder than, the Great Lakes. Data centres, filled with computers, generate heat as they process information. If this heat is not dissipated, it can lead to system failure. Water can act as a coolant by circulating through the system, and cooler climates help reduce the energy needed to maintain these systems.

At the time of reporting, May 2025, nearly one in four of the estimated 3,680 data centres in the US are located in Great Lakes states.[133] Cleveland alone has over twenty data centres within twenty-four kilometres of Lake Erie, with many more expected to come online in the region. Data centre map counts 277 centres in Canada, though other sources suggest the number is closer to 340. The Greater Toronto Area, along the shores of Lake Ontario, is home to 103 data centres.

In April 2024, Amazon announced plans to develop around $11 billion worth of data centre facilities about thirty-two kilometres from Lake Michigan in New Carlisle, Indiana. Google and Meta have also announced plans to invest $2 billion and $800 million, respectively, in their data centre operations in other parts of Indiana. Similar developments are occurring in Great Lakes communities in Illinois, Michigan, and beyond.

In Mount Pleasant, Wisconsin, Microsoft is building data centres on three separate sites near Lake Michigan. Globally, Microsoft's 300 data centres consume more than 125 million litres of water per facility each year, equivalent to 15,000 Olympic-sized swimming pools. While Microsoft's 2024 Environmental Sustainability Report mentions goals related to construction waste and air filtration at its data centres, it does not address water conservation efforts.

A Microsoft spokesperson stated that most of their data centre facilities in Mount Pleasant will not require large quantities of water due to a closed-loop cooling system that uses a combination of chillers and recycled water.

Some experts suggest timing data centre workloads to coincide with the coolest hours of the day. Closed-loop cooling systems, like those proposed for Microsoft's Mount Pleasant facilities, are another way to use water more sustainably. Yet some argue that data centres are not environmentally friendly just because they do not emit smoke.

The use of AI by Amazon, Microsoft, Alphabet (Google), and Meta (Facebook, Instagram, WhatsApp) spiked their emissions by 150 percent between 2020 and 2023, due to the vast amounts of energy required to power data centres, according to a United Nations report. "The rapid growth of AI is driving a sharp rise in global electricity demand, with electricity use by centres increasing four times faster than the overall rise in electricity consumption," the report found.[134] AI impact on climate change, and the parallel impact on our water supply, both quantity and quality, must not, be underestimated.

Helena Volzer, senior source water policy manager at the Alliance for the Great Lakes, expressed concern about the amount of water data centres use, largely because the exact amount is unknown. She noted that less than a third of data centres track their water usage, and there is no requirement for them to do so.

Volzer emphasized that the lack of knowledge, combined with existing water use demands from other water-intensive industries, climate change, and potential population growth driven by economic development, could strain water resources, especially groundwater.

Chapter 39

Water Wars

I suggest that this story from Nanaimo on Vancouver Island is just the beginning of an increasing numbers of water wars.[135]

Across Canada, a new gold rush is underway, and most people have no idea it is happening. The federal government has set aside $700 million to lure data centres with promises of cheap hydroelectric power and a cool climate. At least eight projects are underway to build hyperscale data centres in Canada, according to the federal government. Microsoft has taken the lead in building data centres with AI capacity in Canada. The Washington-based corporation purchased seven large tracts of land in 2021, including a golf course near Québec City and a former department store in the Toronto suburb of Etobicoke.

Towns and cities everywhere will soon face the same challenge Nanaimo residents face now. The big problem is that the investor presentations never mentioned water. Enormous quantities of water.

"There's barely any regulation in place," said Geoff White of Ottawa's Public Interest Advocacy Centre. While communities in the United States, Europe, and Latin America have organized protests and passed restrictions, Canadians remain dangerously uninformed. "If we're racing ahead, thinking only about economic benefits, and not about downstream impacts to our environment, that's negligent."

The numbers are staggering once you know where to look. A 2023 study found that generating just ten to fifty ChatGPT responses consumes about half a litre of water—mostly potable municipal water. Microsoft's Etobicoke facility has been approved to use nearly forty litres per second, roughly 1.2 billion litres annually. One city councillor there hadn't even known about the project until a reporter called him.

This was the pattern Kathryn Barnwell, a retired English professor, discovered: communities blindsided by projects they didn't understand until it was too late. The companies made promises: Microsoft said they'd only draw water when temperatures exceeded 29.4 degrees. But in the Netherlands, their data centre consumed four times the promised amount while local farmers were asked to cut back. When confronted, Microsoft offered vague explanations about estimates being based on "consumption at that time."

Other communities are waking up, though often too late. In Indianapolis, residents organized for months, educating themselves and their neighbours about water consumption. When they finally forced Google to abandon a billion-dollar data centre, the city council chamber erupted in applause that lasted nearly a minute. But for every Indianapolis, there were towns like Etobicoke where the deals were done quietly, with no public opposition.

The opacity was deliberate. Amazon's Varennes facility in Montréal, operated since 2018 without a water metre—nobody knew their consumption. With $100 billion revenue in 2023, they only paid $153 annually for water. Google even paid for a lawsuit against a newspaper seeking consumption figures. When the suit was dropped, the numbers showed Google used a quarter of the city's water.

Nathan Wangusi, who worked on sustainability at Amazon, understands why companies fought transparency. "There's a tendency to deny or cover up that water consumption exists," he said. They knew that informed citizens would ask difficult questions.

Barnwell wanted citizens of Nanaimo and other communities to understand that the companies counted on residents not knowing, not asking, not organizing until the data centres were already built. She'd read about communities across the United States and Europe that accepted data centres only to "regret it really, really deeply," as she put it. The pattern was always the same: promises of jobs and economic growth, followed by water shortages, broken commitments, and the realization that aside from energy regulators, there was virtually no government oversight.

Not every potential data site across Canada has a Kathryn Barnwell to stand up to the tech giants. Small cities and towns are being courted, while either biased, uninformed or misled mayors are dreaming of modernization. Geoff White's warning echoed in her mind: "Our water is highly sought after, and this will increase as the world gets hotter."

Canada is jumping into the AI race with almost no mechanisms to protect its water supply. But it isn't too late—not if people learn what is happening before the deals are signed.

She had a message for Mayor Krog, and for every community facing the same choice: "We have the opportunity here to stop it before it starts."

The mayor talked about jobs and progress, but Barnwell had done her homework, and she knew other residents could do the same. They just needed to know the questions to ask, the promises to doubt, and the courage to say no before it was too late. Indianapolis had shown it was possible. Communities everywhere needed to learn that lesson before the tech giants come calling with their glossy presentations and empty promises.

Nanaimo could have been an example of either resistance or regret. Barnwell did everything in her power to make it the former. Sadly, June Ross of Vancouver Island Water Watch Coalition reported that, "The Data Centre zoning occurred in 2022, and the plan has been in place since 2023. I, like many in the City, had no idea this has been in the works for 3 years, so we have been peddling uphill! It is a done deal and now we will have to monitor as it grows."

CHANGING LAWS FOR E-WASTE

By age sixteen, Alex Lin had already driven major changes to combat e-waste—discarded electronics including computers, phones, and batteries. Inspired by a *Wall Street Journal* article, he initiated action with his community service team.

They recycled 300,000 pounds of e-waste and successfully lobbied Rhode Island's legislature to ban e-waste dumping in 2006.

This led to a 2008 producer responsibility bill that held manufacturers more accountable. Their work expanded globally, helping establish similar teams in the Philippines, Mexico, and Kenya to address local e-waste problems.

Chapter 40

Profit Over People—the Fight Against Bottled Water

Tofino, a city located on Vancouver Island's Pacific coast, has made history as Canada's first municipality to prohibit the sale of single-use plastic water bottles. Mayor Dan Law explained that this decision represents the culmination of the community's longstanding efforts to minimize plastic waste.

"Tofino is making another significant stride in safeguarding our oceans, coastlines, and local wildlife by banning single-use plastic water bottles of one litre or smaller," Law stated. "This regulation demonstrates our collective commitment to environmental stewardship and the well-being of future generations."

Despite a 2019 resolution from the Union of British Columbia Municipalities (UBCM) calling for a halt to groundwater extraction licenses for commercial bottlers, the provincial government has not yet acted.[136]

Recent polling shows growing public concern, with 66 percent of British Columbians worried about potential water crises in their communities, up from 57 percent in 2018. Eighty-five percent oppose bulk water extraction by commercial bottlers.

A University of Victoria Environmental Law Centre report warns that British Columbia's water bottling industry has reached a critical tipping point, with multiple new license applications pending, including two in the province's most arid regions. The report recommends a moratorium on new permits, citing public concerns, insufficient Indigenous consultation and plastic pollution.

Critics highlight the extremely low cost of extraction—just $2.25 per million litres—which generates only $44,000 annually for the province from 16.75 billion litres of licensed water. This volume equals the yearly consumption of over 92,000 residents, while twenty First Nations communities remain under DWAs.

Experts warn that climate change is increasing water scarcity, with 63 percent of British Columbians already living in water-stressed areas. Environmental Law Centre director Deborah Curran emphasizes the need for immediate action, stating that unlimited water use is no longer sustainable given rising droughts and wildfires.

In 2022, Bruce Gibbons and his wife Nicole Poirier learned a neighbour had received provisional approval to extract and bottle 10,000 litres daily from their Merville aquifer.[137] Gibbons believes no one should profit from communal aquifers, a view shared by most British Columbians. A 2018 survey found 91 percent consider fresh water "our most precious resource," while 85 percent view it as a fundamental human right.

His concern was heightened because the area is Agricultural Land Reserve where farmers rely on the aquifer for irrigation. Four years later, he continues advocating despite rejected environmental appeals, unanswered FOI requests and provincial government resistance.

The 2016 Water Sustainability Act aims to ensure sustainable fresh water supplies yet permits licensed commercial extraction. After the Comox Valley Regional District rejected the rezoning application, Gibbons founded the Merville Water Guardians, which grew rapidly as he addressed local governments provincewide. His message resonated: water bottling is unnecessary given British Columbia's safe drinking water, supports fossil fuel industries and creates waste—only 15 percent of plastic packaging is recycled in Canada.

Over a dozen jurisdictions have amended bylaws to prohibit local bottling operations. Despite UBCM passing a resolution in 2019 calling for a halt to extraction licenses, the province has offered only empty rhetoric. Gibbons criticizes the province for offloading responsibility to local governments.

More communities are increasingly joining the battle to protect their water resources, as large corporations move in to take water from natural springs and/or aquifers, leaving local residents uninformed and at a distinct disadvantage. Some bottling companies, including certain Canadian operations run by French-owned Culligan International, source their water from municipal water systems, then purify and market it as treated water. "This concerns certain consumers," notes the Canadian Bottled Water Association's Griswold. "The immediate response is questioning why one would pay for tap water. However, once it undergoes treatment, it's no longer simply tap water."

In an era when politicians rarely demonstrate moral leadership or prioritize the common good, it's somewhat encouraging that one water bottling company briefly chose the ethical path. On July 24, 2025, Diamond Springs Water Bottling Cooperative announced that due to the drought impacting the Fanny Bay community and causing some residential wells to run dry, the worker- members had decided to halt all operations for a minimum of two years.

This pause would allow time to better understand groundwater resources and facilitate broader community dialogue. This exemplifies how things should function: people before profits.

However, it was not a large corporation who took this position, but a co-operative or member/employee-owned business. One can only wish that every community battling corporate bottling facilities could establish cooperatives and achieve such strength and success. Diamond Springs website continues to explain:

"Worker cooperatives represent a fundamental departure from traditional business structures. A worker co-op's main objective in running an enterprise is serving its employees and community, rather than serving capital owners. The aim is creating optimal employment conditions for members while delivering products or services to customers and the community at reasonable prices that fulfill their needs and contribute to community sustainability."

Corporate water extraction affecting local drinking supplies is not confined to British Columbia.[138] A confrontation emerged in Guelph, Ontario, and centred on the contentious matter of Nestlé Waters Canada's application to renew its water extraction permit for the Aberfoyle bottling facility. This permit has become a point of intense controversy among environmental advocates, community officials and business interests following revelations that the province charges bottled water corporations merely $3.71 per million litres extracted.

The matter holds particular significance for Guelph, Canada's largest municipality depending exclusively on groundwater for its water needs. The Aberfoyle extraction draws from the identical aquifer supplying Guelph's water system.

Water protection advocates condemned the province for allowing Nestlé to continue extracting water throughout a summer drought, expressing alarm that such extraction might drain the aquifer upon which Guelph depends. Since summer 2025, the province is examining Nestlé's water extraction permit.

"Many residents are worried that despite current reports showing water flowing from taps adequately, given compelling evidence of recent drought conditions, climate change impacts, plus the provincial Places to Grow Act mandate requiring our city to expand population by 46 percent by 2041, we recognize our increasing water needs and understand groundwater remains our sole source," Gordon explained to CBC News. Some critics have accused Guelph Mayor Cam Guthrie of attempting to suppress discussion on this matter.

This challenge is not limited to Canada. The network of clear streams comprising California's Strawberry Creek run down the side of a steep, rocky mountain in a national forest two hours east of Los Angeles.[139] Last year Nestlé siphoned 45 million gallons of pristine spring water from the creek and bottled it under the Arrowhead Water label.

Though it's on federal land, the Swiss bottled water giant paid the US Forest Service and state practically nothing, and it profited handsomely: Nestlé Waters' 2018 worldwide sales exceeded $7.8 billion.[140]

Conservationists say some creek beds in the area are now bone dry and once-gushing springs have been reduced to mere trickles. The Forest Service recently determined Nestlé's activities left Strawberry Creek "impaired" while "the current water extraction is drying up surface water resources."

Meanwhile, the state is investigating whether Nestlé is illegally drawing from Strawberry Creek and in 2017 advised it to "immediately cease any unauthorized diversions." Still, a year later, the Forest Service approved a new five-year permit that allows Nestlé to continue using federal land to extract water, a decision critics say defies common sense.[141]

TAKING A STAND FOR CLEAN WATER ACCESS IN CANADA

Nestlé extracted an estimated 3.6 million liters of water from Grand River, Ontario's aquifers, reselling it to Indigenous communities at inflated prices while paying only $503.73 per million liters to the provincial government. Residents received no consultation or compensation, forcing them to walk miles for water or buy bottled water.

Six Nations environmental activist Makaśa Looking Horse, twenty-four, led a tireless campaign against Nestlé through legal actions, UN speeches, and demonstrations. In 2018, she organized a massive World Water Week rally.

Her efforts culminated in a cease and desist letter, and in March 2021, Nestlé sold its North American bottled water operations—a victory for activists, though Looking Horse continued advocating for compensation.

Part Four

Urban and Rural Communities at Risk

Chapter 41

Walkerton

Since it began operation in 1987, Well Five in Walkerton, a town in Ontario, experienced sporadic bouts of contamination due to its lack of source water protection and the inadequate treatment and monitoring of water quality. To make matters worse, this information was not revealed to the local health unit. This is why water contamination was not suspected as the cause of community illness until many residents became seriously ill in May of 2000.

Bruce Davidson, one of the founders of Concerned Citizens of Walkerton, received a phone call from a Walkerton resident in late spring of 2000. She told him when she visited her doctor, he ordered several lab tests to assess her condition. After looking at her lab results he asked her if she had recently travelled outside the country—perhaps South America or Africa?

She had not even left Canada. He was surprised because her test results were what he would expect to see from someone who had been drinking untreated water.

Bruce continued: "Many residents, including myself, experienced bouts of painful abdominal cramping and bloody diarrhea months before the drinking water was officially deemed to be contaminated. These often-embarrassing conditions were not something that one would generally discuss with neighbours. Most people assumed it was 'just something they ate.'"

The Walkerton Water Disaster of 2000 clearly illustrates the vital importance of water quality monitoring. Proper monitoring and reporting would allow the medical community to immediately understand the source of communal infections.

One can only wonder how many Canadians have fallen ill or, perhaps, even died, as the result of drinking contaminated water which had gone undetected.

Following the Walkerton tragedy, Dennis O'Connor, Associate Chief Justice of Ontario, acted as commissioner for what was known as the Walkerton Inquiry. After a scientific advisory panel assessed what happened, the Inquiry made 121 recommendations on a wide range of areas related to protecting drinking water.

These recommendations became the building blocks of Ontario's drinking water protection framework implemented by the Ontario government.[142] They include the Clean Water Act, established in 2006. Under this legislation, nineteen local multi-stakeholder source protection committees were established, guiding source water protection efforts in areas across Ontario.

Today, Ontario has a comprehensive drinking water safety net from source to tap. The safety net includes strong legislation, stringent standards, regular and reliable testing, highly trained, certified operators, licensing of drinking water systems, regular inspections of drinking water systems and labs that test drinking water, public reporting and a comprehensive drinking water source protection program. Ontarians can be confident that they enjoy clean and safe drinking water, and that the province will continue to take action to safeguard their water and consequently their health. Well, that is what the government would like Ontarians to believe.

While Ontario may brag they have the tightest drinking water guidelines in Canada, one just needs to google Ontario DWAs to see what is really happening. This also reaffirms why it is not feasible to download full responsibility to local municipal or community water treatment operators. The North Bay Parry Sound District Health Unit issued a boil water advisory for users of the Thorne Drinking Water System due to a loss of disinfection that occurred on July 30, 2025.

And it is not necessarily communities of 5,000 population or less. Shelbourne is a population of 9,000 people: the BWA issued in Shelburne affected residents due to water main repairs. Or the Simcoe Muskoka District health unit declared water quality in the areas of Brechin and Lagoon City may be compromised due to elevated turbidity levels in the raw water supply. If turbidity levels continue to rise, a boil water advisory will be put in place. If you still think this is just a few places that don't affect you, check out this website, look at DWAs for any specific area. As of this writing in December 2025, there are 181 DWAs in Ontario.[143]

Sadly, all the improvements that resulted from the Walkerton disaster do not apply to First Nations communities in Ontario, because First Nations fall under federal jurisdiction. Ontario residents, under provincial jurisdiction, now enjoy robust water safety protection, but Canadians in other provinces do not have equivalent safeguards. Canada's constitutional division of powers places drinking water under provincial jurisdiction, requiring each province to develop its own guidelines independently.

The critical question remains: Why haven't other provinces proactively adopted similar comprehensive measures? Must each province wait for its own devastating public health crisis? The benefits are clear. You just have to check the number of BWAs at any given time and compare the number of them in effect in Ontario to the number of them in effect in other provinces.

Try asking your doctor to test you for *E. coli* because you have had stomach issues. I expect their response will be: "There's no need to test unless you've been out of the country."

We need massive education for doctors to fully understand waterborne illnesses. They need to realize we can experience developing countries' illnesses in our own backyards.

The advisory panel of the Walkerton Inquiry is a list of extremely well published and highly respected scientists with international reputations. Prof. George E. Connell, OC, FCIC, FRSC, a biochemist;

Steve E. Hrudey, a professor of environmental health scientist who wrote many peer-reviewed papers with Dr. H_2O; Prof. William Leiss, president of the Royal Society of Canada; Douglas Macdonald, PhD, a lecturer in environmental studies; Dr. Allison J. McGeer, a specialist in infectious diseases; and Prof. Michèle Prévost, an internationally recognized expert in environmental engineering.

I estimate these highly accomplished individuals have published hundreds of scientific papers in total. What I cannot find is how even one of those papers has resulted in positive change toward providing Canadians with truly safe drinking water. How has their outstanding science been applied to affect change? Are their papers collecting dust on academia bookshelves? Billions of taxpayers' dollars are spent by governments of all levels, under a variety of pseudonyms, all appeasing science. I don't understand why the government appears to be ignoring the indicators of their research or failing to fund applicable existing research which will improve water treatment processes, and above all, protect our source waters.

Canada is a prosperous nation with enough resources for everyone. Yet some have excessive wealth, while many struggle to survive. Too many people don't have access to the basic human right of safe drinking water. Don't think that if you live in cities or towns in Canada your water is safe. You may have access to what you believe to be safe drinking water, but you must ask the million-dollar question: Is it really safe?

PADDLING VS. PUBLISHING

Earthviews founder, Brian Footen, a former fisheries scientist turned entrepreneur, paddled for conservation using a 360-degree GoPro Max camera and fishing kayak. He created immersive digital maps of vulnerable waterways to engage the public, collect scientific data, and document regions affected by development and climate change.

Footen paddled ten to fifteen miles daily near shore, capturing panoramic images every ten seconds while logging water quality data and animal sightings.

He first documented Puget Sound, then surveyed the Great Salt Lake, Lake Tahoe, Lake Mead, Lake Powell, and Rio Grande during the historic American West drought.

Footen noted that previously, "I published research papers, and they were read in journals by my colleagues and then put on a shelf somewhere. Rarely did they move policy."

Chapter 42

Well Well

Many rural communities and farmers in Canada struggle to get safe drinking water from wells.[144]

Manitoba Environment and Climate Change recommend testing well water at least once a year, especially after snowmelt, flooding, heavy rainstorms or if there is a change in water colour, clarity, odour or taste. The Private Well Testing Program offers a partially subsidized annual test to detect total coliforms and *E. coli* in well and cistern water.

I imagine Manitoba Environment feels this is a reassuring program to offer rural residents. In *Water Confidential,* I explained how coliforms and *E. coli* were the perpetrators both in Walkerton and at six First Nations communities. I also shared how Dr. Fricker, a water quality specialist from the UK, was often contracted to fix water treatment plants around the world. He was the lead adviser during the cryptosporidium outbreak in Sydney, Australia. I coordinated a conference in Saskatoon in 2004, "The Future of Water Treatment in Canada," where Dr. Fricker began his presentation with a question: "If you have no coliforms or *E. coli* in your drinking water, what does it mean?"[145] He posited the answer: "It means that you have met regulatory requirements. Does it mean that you have produced safe drinking water? Absolutely not! If you only take one thing from my presentation today, let it be that testing for guidelines does nothing to protect public health."

Silence blanketed the large conference room following his assessment: Canada's safe drinking water guidelines were effectively futile. Of course, it is good to know if *E. coli* or coliforms are present in well water, and then to remove them. But there are so many more compounds that should also be tested for. As many rural communities are based on agriculture, and have wells, I think pesticides would be a good place to start testing.

It seems to me that Manitoba Environment and Climate Change have not read my book or followed the world leading expert on water quality, Dr. Colin Fricker. They do offer that if the test for coliforms or *E. coli* is positive, the homeowner is notified and instructed how to disinfect the water source. A free coupon to re-test after treating the well or cistern is also issued.[146]

Does that make you feel secure that farmers with wells have safe drinking water? Farmers do not pay any water bills, so do not have any legal entitlement to services. The farmers could send their water to an accredited laboratory if they wanted more comprehensive test results. This is a matter of human health, where all Canadians are entitled to truly safe drinking water. I feel this is another example of if you don't look for something you can say you didn't find it. Apparently, Manitoba Environment and Climate Change hasn't been taught to inquire aggressively. If they were to look for pesticides in the tap water of these rural farmers, I suspect they would be highly likely to find them.

The Canadian Guidelines for drinking water quality (and I believe the provincial standards/regulations) have each pesticide listed separately. Whereas in Europe, there is a total allowable maximum for pesticides. This is because there are cumulative effects. If you add the amount of pesticide of each type that are allowed in water in Canada, it is extremely high compared to the total allowed in the European Regulations.

November 2025 the federal government proposed to eliminate

mandatory safety reviews for dangerous pesticides. These reviews, called re-evaluations, ensure pesticides continue to meet health and environmental standards as new science emerges. Without them, unsafe products could remain on the market indefinitely.

Re-evaluations were added to law in 2002, and courts have identified them as essential for public safety. Despite years of public consultation on improving pesticide oversight, this change was quietly inserted into the 2025 federal budget.

Combined with another proposal to eliminate the current five-year pesticide registration renewal process, these changes would dramatically reduce oversight of pesticide safety, potentially putting Canadians and the environment at risk.[147]

PREVENT CANCER NOW – MEDIA RELEASE

A coalition of nearly thirty organizations demands the federal government halt plans to weaken Canada's pesticide regulations and instead launch overdue comprehensive reform. The government intends to abolish cyclical pesticide re-evaluations, further empowering an industry-friendly regulator that has lost public trust.

The Pest Management Regulatory Agency spent $42 million on a "transformation agenda" with no improvements. The legally required parliamentary review of the Pest Control Products Act is four years overdue—last conducted in 2015.

"No modification should be made without full understanding of current science and open debate involving all stakeholders, not only the pesticide industry," warns the Coalition. This demands accountability: transparent public process over industry influence in decisions affecting farmers, health, and environment.

Chapter 43

Pesticides

A recent study by Sébastien Sauvé, an environmental chemistry professor at Université de Montréal, revealed that various pesticides can still be found in tap water even after treatment at a water plant.[148] The study was published in the June issue of the scientific journal *Water Research*.

From 2021 to 2023, researchers collected water samples twice a week from the Châteauguay River in Québec's Montérégie region.[149] They compared samples taken from the river as water was being pumped into a treatment plant, with samples of treated water leaving the plant for distribution as drinking water. The river was chosen due to its location in areas of intensive agricultural activity, where pesticides used by farmers eventually make their way from the fields into waterways.

The study aimed to observe how pesticide concentrations in the water varied over time, both before and after processing at a treatment plant. The highest concentrations of pesticides were found in June and July at the beginning of the growing season. Although none of the samples exceeded Québec guidelines for safe drinking water, researchers discovered that the treatment process did not effectively filter out pesticides.

I believe it is worthy to note that Québec does have regulations for their drinking water under the LQE (Loi sur la qualité de l'environnement).[150] Like all regulations, they are useless unless enforced. While the LQE governs environmental protection, including drinking water, it doesn't directly lay charges for non-compliance. Instead, the regulation respecting the quality of drinking water outlines specific requirements for drinking water quality and the consequences for not meeting them. These consequences can include notifying the authorities and the public, and potentially other measures, but not necessarily formal charges.

This study was conducted where, I presume, they had an effective water treatment system—but then I wonder if they looked at the efficacy of that system by testing other compounds. I am sure there are hundreds more water treatment systems across the country which are ineffective, and no one appears to be testing them.

At least fifty different types of pesticides and their breakdown products, known as metabolites, were detected in tap water, sometimes in higher concentrations than in the source water. I'm sure that would be over the total European limit because they would add them all up!

Metabolites are molecules formed from the degradation of pesticides. "There is quite a bit," Sauvé said, noting that some pesticides were found in nearly all samples, while others were less frequently detected. "One of the main ones, or the highest concentration, was for glyphosate," Sauvé said, referring to the primary active ingredient in many herbicides, commonly sold under the brand name Roundup, which is used to kill weeds.

In Quebec (and many other provinces), many municipalities, including Montréal, have banned the sale of glyphosate or Roundup for domestic purposes over concerns for the environment and possible impact on human health. I would like you to remember two things: first the name glyphosate, which I will re-visit in later chapters; second, I wonder what *quite a bit* amounts to. I like sugar in my coffee, but a double-double is way too sweet, even for me. Do I use quite a bit of sugar in my coffee or does Tim's? Just how much is *quite a bit*? For me, just a teeny-weeny tiny bit of pesticide in my drinking water is unacceptable.

Why is Canada not banning glyphosate on a national basis for this highly dangerous and corporately protected monster. In 2015, WHO and the International Agency for Research on Cancer concluded that glyphosate was probably carcinogenic to humans.[151]

Sauvé expressed concern about the combined effects of various pesticides rather than the impact of any single one.[152] He explained that the mixture of different pesticides creates a cumulative effect. The interaction of these compounds can influence toxicity, potentially adding up, partially neutralizing each other or even multiplying. With around fifty different pesticides and metabolites measured, Sauvé emphasized the uncertainty surrounding the combined toxicity of these compounds and highlighted the many unknowns in such situations.

I compare this situation to taking too many medications: you need a caring pharmacist who can keep watch and ensure you don't take conflicting or interacting medications. This is where Europe is far ahead of Canada, and where our federal government is failing in protecting the health of Canadians.

Maryse Bouchard, a professor of environmental health at the Institut national de la recherche scientifique, told Radio-Canada that Suave's study on temporal trends of forty-six pesticides and eight transformation products in surface and drinking water in Québec underscored the importance of better evaluating the risks associated with the "cocktail effect."[153]

She found it alarming to see the number of different pesticide molecules, many of which have known toxicity. Bouchard's research focuses on the effects of various environmental contaminants on community health.[154] Her work aims to identify and quantify health risks through epidemiological studies, including risks to the nervous and endocrine systems. She has conducted research on the risks associated with exposure to metals like manganese and lead, as well as pesticides and other chemical products. Many of her studies focus, or have focused, on population groups with vulnerability factors like fetuses and children, seniors, Indigenous peoples and people who work with toxic products.

While Sauvé mentioned that there is no immediate cause for alarm for tap water consumers, the findings do raise some concerns.[155] He suggested that instead of panicking, people should ask their local authorities if improvements can be made to the water system. Of course that is what he said. Don't panic! I remember too well the far-reaching politics of Roundup. That was why Dr. H_2O was bound and gagged by the high ups at the Saskatchewan Research Council (SRC) decades ago. Scientists must be allowed to voice their sound uncompromising opinions based on the science they conduct.

What he'd like to see are stricter guidelines similar to those used in Europe. Specifically, he'd like to see an upper limit on the total concentration of combined pesticides allowed in tap water.[156] We must demand national drinking water regulations and raise the bar.

REJECTING CYCLICAL PESTICIDE RE-EVALUATIONS

Health and environmental organizations with expertise in Canadian pesticide regulation under the Pest Control Products Act oppose the government's Budget 2025 proposal to eliminate mandatory fifteen-year cyclical re-evaluations.

These re-evaluations are a critical pillar ensuring pesticide health and environmental risks remain acceptable, as recognized by the Federal Court of Appeal. They provide comprehensive review, public participation, and mandatory checkpoints supplementing the Minister's discretionary powers, unlike limited-scope special reviews.

The proposal is alarming alongside plans to eliminate five-year validity periods, creating indefinite registrations without renewal checkpoints. Together, these changes would make post-market pesticide risk assessment almost entirely discretionary, significantly weakening public health and environmental protections that organizations—not government—are fighting to preserve.

Chapter 44

Rural Canada

We stopped in Princeton, British Columbia, where I used the washroom in the tourist office and noticed a sign on the wall that read: "Do Not Drink the Water Here." I asked the woman behind the counter why.

"It's almost three years now. The old system couldn't be fixed, and they are building a new water treatment plant, which should be completed soon. We don't worry about it anymore, it's been so long, and none of us living here get sick and we drink it all the time."

I told her about my friend who almost died from *H. pylori* and stressed how many waterborne illnesses take decades before they present as health issues.

"Do you still think it's okay to drink the water here or to tell visitors they can drink it?"

Judging by the change in colour of her face and the look in her eyes, I think she got the message. I hope so.

I walked to the local bookstore and introduced myself, hoping the owner would stock my memoir *Water Confidential*. "No, we don't need it. Everyone here is sick of reading about drinking water."

I doubt he realized he used a pun. Quite possibly the people of

Princeton would literally become sick in years to come, but not from reading about water. We decided we didn't want to drink our coffee in Princeton and continued driving.

AMERICA'S ONLY FULL-TIME INDUSTRIAL-STRENGTH RIVER CLEAN-UP ORGANIZATION

What began as one person cleaning one river evolved into an industrial-strength organization with over 120,000 volunteers cleaning America's major rivers. Founded by Chad Pregracke, who grew up near the Mississippi River, Living Lands and Waters started in 1997 when he single-handedly removed 45,000 pounds of trash.

Over twenty-five years, the organization worked on twenty-five rivers across twenty-one states, removing over 11 million pounds of trash. A full-time crew lived on a house barge, traveling nine months yearly with extensive equipment, hosting cleanups, tree plantings, and educational workshops. Following the December 2021 Kentucky tornado, they removed 1.2 million pounds of debris from Kentucky Lake in their largest single cleanup effort.

Part Five

Is Any Drinking Water Safe In Canada?

Chapter 45

When Contaminants Multiply

In the summer of 2025, there were thirty-five DWAs in First Nations communities in Canada.[157]

At the same time, there were over 1,400 DWAs in non-Indigenous communities with British Columbia leading the way with 448![158] This is nothing new. In 2001, there were 304 communities under BWAs in British Columbia. I have been unable to determine how many communities in British Columbia do not treat their drinking water; I suspect it is not something those in high places want to brag about. I assume it is simply the result of having once had the most pristine water supply in Canada.

As the number of DWAs steadily increases, I see this as an indicator of risk, of the likelihood that eventually untreated or ineffectively treated water is going to make people critically ill. DWAs are the canaries in the water treatment plants.

British Columbia has fewer problems with water pollution compared to other provinces, but the lower Fraser River Basin has high levels of agricultural runoff and municipal wastewater discharge, which leads to nutrient enrichment.[159] About 90 percent of the province's municipal wastewater is discharged into the lower Fraser River or its tributaries

Susan Blacklin

There are literally hundreds of small towns in Canada with DWAs, and they are called for a reason. I would like you to note the message here, taken from the National Collaborating Centres for Public Health:

"Canada doesn't keep complete or consistent records about drinking water problems and waterborne disease outbreaks. If all provinces tracked this information the same way, it would help create better water safety rules. Most waterborne disease outbreaks in Canada happen in small water systems that serve 5,000 people or fewer. These outbreaks usually happen when several things go wrong at once. Common problems include not protecting the water source (like lakes or rivers) and not treating the water properly before people drink it." 160

Research like this should influence politicians' agendas. It should be obvious; we need to protect source waters. We need to ban the release of toxic effluent by industry and agriculture, which is the single most critical and cost-effective means to effectively produce truly safe drinking water.

Analyses suggest small drinking water systems face challenges associated with infrastructure, technology and financial constraints. Investments in drinking water systems and operator training have the potential to reduce the burden of waterborne disease in Canada.

I do not believe there are financial restraints limiting improved water quality for Canadians, rather there is massive misappropriation of government funds. The second most critical factor following protection of source waters must be Dr. H_2O's theory—and now mine —that training WTOs to work on ineffective water treatment systems is futile and further misuse of taxpayer money. You can think of it as training people to repair cell phones and they return home to use old wall phones with party lines. Now this statement will confuse younger readers!

According to the Conference Board of Canada, Canada's water quality is at risk from poorly treated municipal waste, industrial effluent and fertilizer runoff from agriculture.[161] Most of the nitrogen

232

and phosphorus released into the environment comes from these three sources. They actually state: Although toxic effluents are heavily regulated in Canada, release of nutrients into the watershed is common.

I think that the person who wrote this probably had legal advice or perhaps even legal dictation. I could not find an example of toxic effluents being regulated, and they certainly are not heavily regulated. I find this rather cynical, being as effluent is allowed to be released into our waterways. Once in the water, it seems the response is: "oops, a little too much here, now we'll just scoop some out." That isn't going to work; once in the water, it is there to stay.

As I have shared, pulp and paper mills, diamond mines, agriculture and the oil fields (to name a few) produce scarily large amounts of toxic effluent, and there isn't anything anyone reliant on drinking the water can do about it. We need to ban the application of known toxic chemicals as we cannot contain how they leach and penetrate water sources. Similarly, industry should be held accountable for financing effective water treatment plants that can remove the toxic elements they have dumped into waterways.

Scientists studied forty-six pesticides and eight breakdown products (what pesticides turn into over time) in Québec's water from 2021 to 2023.[162] They tested both river water and drinking water twice a week for 838 days at a water treatment plant. Their goal was to see how much of these chemicals were in the water and whether they posed health risks to people drinking the water.

What they found:

- Pesticide levels were highest in June and July (summer months)
- Some breakdown products had higher levels in the water than the original pesticides
- Water treatment didn't significantly reduce the total amount of pesticides and breakdown products
- Surprisingly, eleven pesticides and 1 breakdown product were found at higher levels in drinking water than in river water

- Health risks were eighteen times higher during summer than the rest of the year
- Breakdown products posed 1.4 to 144 times more health risk than the original pesticides

Why this matters: This was the first long-term study to compare the health risks of pesticide breakdown products versus original pesticides in drinking water. The results show that Canada urgently needs regulations to control these breakdown products in drinking water, not just the original pesticides. We need to adopt the regulations in the UK, for the cumulative effect of pesticides, but also the need for drinking water regulations here in Canada.

But it's not just parasites and viruses and bacteria that you need to worry about in your water. It is also the quantity of chemicals being used and applied with the sound approval of our federal government.

CLEAN WATER ACTION (USA)

Since its founding during the 1972 Clean Water Act campaign, Clean Water Action worked to win strong health and environmental protections through issue expertise, solution-oriented thinking, and people power. The organization's mission focused on protecting environment, health, economic well-being, and community quality of life by organizing grassroots groups, delivering successful campaigns, and electing environmental candidates at all government levels.

Clean Water Action took action to protect drinking water, environmental waters, and health through campaigns that removed health-harming toxics from everyday products, protected water from dirty energy threats like drilling and fracking, built a future of clean air, water, and energy, and kept laws strong and effective for water and health protection.

Chapter 46

Per- and Polyfluoroalkyl Substances (PFAS) in Drinking Water

In *Water Confidential,* I shared a story about a farmer who could ignite the water coming out of his kitchen faucet, which I suggested might be due to fracking.

Fracking, or hydraulic fracturing, involves injecting pressurized sand, water and chemicals into wells and underground lines to extract oil and gas. This technique is banned in many EU countries, such as Germany, France and Spain, as well as in Australia, but it is prevalent in North America, particularly in British Columbia, Alberta and in the USA in North Dakota, Pennsylvania and Texas.

Research has found that people living near fracking sites face serious health problems. Children in these areas have higher rates of a deadly cancer called acute leukaemia. Pregnant women near fracking operations are more likely to have babies born too early or at dangerously low weights. These early-life problems can cause health issues that last a person's entire life.[163]

A group of doctors in Colorado called Physicians for Social Responsibility investigated what chemicals oil and gas companies were using. What they discovered was alarming: companies were breaking the law by not reporting their chemicals.

Colorado had a law requiring companies to:

- Tell the public what chemicals they use
- Stop using dangerous "forever chemicals" called PFAS (these never break down in the environment or human body)

Despite this law, the doctors found that in just twenty-one months (less than two years), companies injected at least thirty million pounds of unreported chemicals into 675 wells. That's like dumping 15,000 cars worth of mystery chemicals underground, and nobody knows exactly what those chemicals were or whether they included the banned PFAS.

Families living near these sites have no way of knowing what toxic substances they're being exposed to, even though laws are supposed to protect them by sharing this information.

In Canada, there is only one public website where people can find information about chemicals used in fracking: www.fracfocus.ca. This website is run by provincial and territorial governments and regulatory agencies, including the British Columbia Oil and Gas Commission.

However, companies choose voluntarily whether to report their chemicals—they are not required to do so. The information available has many gaps.

At least forty-eight chemicals on Canada's National Pollutant Release Inventory (NPRI) (a list of known pollutants) are used in fracking operations. Many fracking companies label their chemical mixtures as "trade secrets," which means they don't have to tell the public what's actually in them. This leaves Canadians in the dark about what chemicals are being injected into the ground near their communities.

Do you remember Sébastien Sauvé in the chapter "Well Well"? He's the scientist conducting cutting-edge research on pesticides in drinking water. Sauvé is a scientist who speaks the truth and can't be muzzled. My mind recalls his research showing how pesticides interact and become far more threatening in groups. I can't help but wonder if a scientist has conducted research on the effect of combinations of chemicals used in fracking and their interactions, or threats, when applied in various configurations.

Decades ago, I shared a recipe for bread. It asked for one package of yeast. My friend had brought a block of yeast from her local bakery and she put the entire block in the recipe; it took her a week to clean the exploded gobs of dough from the inside of her oven. How chemicals interact and the amounts used are critically important.

My main concern is the impact of fracking on our water. In 2022, Sauvé and his research team launched a major study testing drinking water for PFAS at nearly 400 locations across Québec.[164] They collected water samples from public sources, including restaurants, without asking for permission. Of the 500 samples collected, only two showed *no* trace of PFAS. Sauvé shared his data with Quebéc's environment ministry while Health Canada proposed new national guidelines to limit PFAS in drinking water. In Quebéc municipalities, PFAS levels were two or three.

Despite the harmful health effects of PFAS, the official response from various levels of government has been muted. In Québec, there is still no enforceable standard for PFAS in drinking water, and Health Canada's guidelines are not binding, so municipalities are not obligated to test their water.

In 2024, the Québec government launched public consultations to regulate PFAS in sewage sludge, making Quebéc the first place in the world to regulate PFAS in sludge before monitoring them in drinking water.

Reaction in affected areas has been mixed, with some officials thanking Sauvé for informing them before it hit the news, while others were furious, telling him it was none of his business. I hope more scientists start to speak out and make their science applicable instead of letting it collect dust on bookshelves. His advice to step out of your comfort zone applies to all Canadians, not just scientists. Together, we can make a difference.

Since then, Sauvé has made numerous media appearances, addressed parliamentary committees in Ottawa and France, and worked with local residents to analyze water and soil samples. He believes public pressure is the only way governments will act, which is why he has gone public with his findings.

Sauvé continues to research PFAS, including in grocery items, and manages a daily stream of media requests for his expertise. He encourages researchers to go beyond publishing their findings and to engage with the media, the public and decision makers to effect change.

The addition of PFAS to Canada's list of toxic substances is unlikely to have an immediate impact on already contaminated communities like North Bay, Ontario, whose water treatment plant on Trout Lake, built in 2010, is not equipped to filter PFAS. They are not alone. I suspect the majority of small-town water treatment plants are in the same position.

In early March 2025, ECCC announced its intention to add PFAS to the country's official toxic substances list. Cassie Barker, the senior program manager at Environmental Defence for toxics, stated that certain types of PFAS can harm the kidneys and immune system, "It harms the endocrine system," she said. "The basic building blocks of your body are being interfered with by very small levels of PFAS."

Ontario's Ministry of the Environment has an interim advice value for PFAS in drinking water set at seventy nanograms per litre, equivalent to 3.5 drops in an Olympic-sized swimming pool. Without national laws or regulations, I see no benefit to this piece of paper.

Canada has big plans for its military future, with the federal government pledging to increase defence spending and hire thousands of people in the next decade. The Department of National Defence has promised 668 new residential units on Canadian Armed Forces bases across the country. Freelance reporter Leah Borts-Kuperman has been investigating contamination at the Canadian Forces base in Moose Jaw, Saskatchewan.[165]

The slogan for the city of Moose Jaw is "Canada's Most Notorious City," stemming from historical connections to Al Capone. The Canadian Armed Forces base nearby, the home of the Snowbirds, is surrounded by a patchwork of quintessential Saskatchewan cropland and homes. As one of the city's major employers, the base employs about 1,000 active military service members and federal public servants. Over the last year, some of them have begun to testify that they believe the workplace they have dedicated their lives to is making them very sick—in some cases, terminally so.

These workers have attempted to get answers from their employer, the Department of National Defence, but say they have faced skepticism, criticism and retribution for asking questions and speaking out. Still, they are determined to fight for accountability.

Moose Jaw can now claim a completely new interpretation of becoming the Notorious City, as current and former employees have been drawing attention to a rash of cancer cases and other illnesses. The planned new housing is on bases listed on the public inventory of contaminated sites owned by the federal government, which includes thousands of places across Canada where federal projects and activities have left PFAS, PCBs and other toxins in the water and soil.

Studies have found PFAS worldwide at very low levels in just about everyone's blood. Higher blood levels have been found in communities where local water supplies have been contaminated by PFAS. People exposed to PFAS in the workplace can have levels many times higher.

PFAS are linked to various cancers, heart issues, immune dysfunction and problems with fertility, pregnancy and child development. While reports are available suggesting that PFAS exposure is associated with an increased risk of testicular cancer and kidney cancer and may be associated with an increased risk of prostate cancer, there are many other scientific reports calling for further research. I suspect that identifying which cancers are caused by PFAS is in early stages. There are numerous compounds in PFAS and the data on the identity, composition and quantity of PFAS used in products and processes are often treated as confidential business information, hampering efforts to estimate exposure sources and routes.[166]

PFAS exposure has been linked to a number of adverse health effects including certain cancers, thyroid dysfunction, changes in cholesterol and small reductions in birth weight.[167] A report recommends that the Centres for Disease Control and Prevention (CDC) update its clinical guidance to advise clinicians to offer PFAS blood testing to patients who are likely to have a history of elevated exposure, such as those with occupational exposures or those who live in areas known to be contaminated.

If testing reveals PFAS levels associated with an increased risk of adverse effects, patients should receive regular screenings and monitoring for these and other health impacts. Guidance on PFAS exposure, testing and clinical follow-up recommends that the CDC, Agency for Toxic Substances and Disease Registry and public health departments support clinicians by creating educational materials on PFAS exposure, potential health effects, the limitations of testing and the benefits and harms of testing. I could not find similar recommended actions in Canada.

PFAS experts, industry representatives, health officials and academics gathered on September 16, 2025, at Toronto Metropolitan University to address PFAS contamination.[168]

Stephanie McFadyen, manager of Health Canada's water quality program, outlined the government's role in setting drinking water objectives. While Canada does not yet have formal guidelines for PFAS in drinking water, it has set an interim objective to reduce exposure.

Does this make you feel as though your health is in good hands? Do you see setting drinking water objectives as a means to protect people's health? You can check various products for PFAS at canopyplanet.org.

A working group is renewing calls to the federal government to add fracking chemicals to the NPRI to advance communities' right to know. Currently fracking is considered part of the oil and gas extraction process, which is exempt from NPRI reporting. The expansion of the industry has concentrated wells in northeastern British Columbia and northwestern Alberta, often close to residential neighbourhoods. Proposed LNG plants on the British Columbia coast could soon increase that concentration extensively.[169]

SHANNEN'S DREAM

Shannen Koostachin was a young Cree education activist from Attawapiskat First Nation, Ontario, who attended classes in portable trailers after a 1970s gas leak polluted her school.

Realizing the government underfunded Indigenous schools—many lacking permanent teachers or safe drinking water—Shannen advocated for "safe and comfy schools," using social media to encourage letter-writing campaigns.

In 2008, politician Chuck Strahl wrote that the government couldn't fund a new school. Shannen's Grade 8 class cancelled their graduation trip to visit Ottawa instead. Confronting Strahl, she said, "I wish my classroom looked like this," referencing his fancy office, and vowed not to quit.

The government promised a new school in 2009.

Tragically, Shannen died in a car accident before turning sixteen. Her legacy lived on through Shannen's Dream organization and Kattawapiskak Elementary School, which opened in 2014.

Chapter 47

Forest Fires

We all know we need water to fight fires, but what about how fires impact our water?[170]

Forever chemicals have polluted the water supply of 60,000 people, threatening human health, wildlife and the wider ecosystem. But activists say this is just the tip of the PFAS iceberg. The quiet French commune of Buschwiller in Saint-Louis is near the Swiss city of Basel. Perched on a hill not far from the Swiss and German borders, it seems like a safe place to raise a child: spacious houses are surrounded by manicured gardens, framed by the wild Jura mountains.

People there use tap water every day, for drinking, brushing their teeth, showering, cooking and washing vegetables. Now, they have learned that chemicals they had never heard of were lurking in their bodies, on their skin, and potentially harming their children. People there are scared, "Even if we stop drinking it, we will still be exposed to it, and we can't really do anything." A letter dropped through the front door of Buschwiller residents from the local authority declared their drinking water was prohibited.

Saint-Louis is now the site of France's biggest ever ban on drinking tap water. Its at-risk residents will rely on bottled water until at least the end of 2025, when authorities hope water filtration systems will be installed. Tests of the local tap water showed levels of PFAS—(forever chemicals) linked to cancer, immune dysfunction and reproductive issues—had reached four times the recommended limit. Shelves were stripped bare as families scrambled to stockpile bottles of water to protect loved ones.

The source was a firefighting foam used at the airport since the 1960s, ending only in 2017, according to the joint statement from the local authority and regional health agency. Toxic residues from the foam lingered, filtering through the soil into drinking water and people's bodies, probably over decades.

"PFAS are not a fire service problem. It's a global problem," said Bryan Ormond, associate professor at the Textile Protection and Comfort Centre in the Textile Engineering, Chemistry, & Science Department at Wilson College of Textiles; North Carolina (NC) State University. "They've just been used in way too many places and too many products. Are there any people on the planet that don't have PFAS? Babies are born with PFAS already in their blood."

PFAS can have a tendency to bio magnify, which means they accumulate in larger levels the higher up the food chain you go. This was reported in the Canadian government's updated draft on the state of PFAS, issued in July 2025, that is intended to guide the decision making on PFAS as a class in Canada.

PFAS surfactants and polymers have many practical applications which has made them a suitable ingredient in Class B aqueous film-forming foam (AFFF) and alcohol-resistant AFFF used for class B fires, like oil, diesel and alcohol fires since the 1960s.

Canada has "hot spots" where higher levels of PFAS are measured in areas where AFFF foam was used to fight fires or for training and equipment maintenance at airports and military buildings. However, contamination is found throughout Canada, not just these concentrated areas.

There has been an ongoing phase out of PFAS in foams, with regulations currently being revised under the proposed Prohibition of Certain Toxic Substance Regulations 2022. Implemented by the Canadian government, these regulations will further restrict any exemptions to the point where there would be a phase-out of the use of AFFF containing PFOA and/or LC-PFCAs as early as 2025."

PFAS molecules are still used in some Class B applications today, but over the last ten years, the need for this material is slowly going away as technology continues to improve. Misconceptions, however, continue about which foams have PFAS, and which do not. It is always best to consult your foam manufacturer, said Mark Biernat, president of Biernat Fire Feu Inc.

In the early 2000s, the Dundurn military base was located a short distance from our farm in Saskatchewan. It had a training exercise where they started a grass fire that quickly got out of hand. Firefighters arrived and put it out. Weeks later farmers in the area learned the foam they had used had impacted their farming. Organic farmers lost their organic license. I wonder what effect it had on the ground water. Or, since the affected land was close to the South Saskatchewan River, which supplies the city of Saskatoon with drinking water, how far did those PFAS reach? What about similar applications in other places? Like Lytton or Jasper, I tried to determine where PFAS were used to fight fires in Canada with no success.

WATER ADVOCATE

Scott Harrison transformed from a New York nightclub promoter into a water advocate after spending two years on a Liberian hospital ship, where he witnessed the devastating effects of contaminated water.

In 2006, he founded Charity: Water, a nonprofit dedicated to providing clean drinking water in developing countries. The organization has achieved remarkable success, mobilizing over one million global supporters to raise $496 million.

These funds have financed more than 111,796 water projects across twenty-nine countries, bringing safe drinking water to communities in need and demonstrating how one person's life-changing experience can create widespread humanitarian impact.

Chapter 48

The Great Lakes

Lake Superior is in the news again. Environmentalists warn that an aging pipeline carrying oil along the bottom of the ecologically sensitive and turbulent Straits of Mackinac, where Lake Michigan and Lake Huron meet, is in such a state of disrepair it could burst at any moment and cause catastrophic damage to the Great Lakes.[171]

Line 5, a 1,000-kilometre-long pipeline owned by Calgary-based Enbridge, carries up to 540,000 barrels of oil and natural gas liquids a day from Wisconsin to Sarnia, Ont., where it is shipped to other refineries in Ontario and Québec. Originally built to last fifty years, it is now over seventy years old and has already spilled more than six million liters of fossil fuels into the Great Lakes Basin. It's at the centre of a politically charged dispute between Michigan Gov. Gretchen Whitmer, who's ordered what she calls the "ticking time bomb" to be shut down, and Canadian officials, including Ontario Premier Doug Ford, who've sided with Enbridge in insisting it's safe to keep running.

The US Army Corps of Engineers has decided to fast-track permits for building a protective tunnel around the aging Enbridge oil pipeline, stoking environmentalists' fears that the project will escape scrutiny, damage the sensitive region and perpetuate fossil fuel use.[172] The move comes after President Donald Trump issued an executive order in January 2025 declaring the US has become too dependent on foreign energy sources. The order mandates that federal agencies identify energy infrastructure projects for expedited emergency permitting from the US Army Corps of Engineers and the EPA.

This tunnel is intended to house its aging Line 5 pipeline; its continued operation threatens the Great Lakes. Instead of decommissioning this old pipeline, Enbridge wants to invest billions into a first-of-its-kind underwater tunnel. This project has never been attempted anywhere in the world, and it could have devastating consequences for the Great Lakes.

Michigan decision makers are currently evaluating whether to grant Enbridge a permit for this tunnel. Enbridge reported earnings of $2.18 billion for the quarter ending June 30, 2025, up from $1.85 billion in the same quarter last year.[173] This makes me wonder what amount of security, or collateral, Enbridge can offer if the leaders in the company are so convinced of this tunnel's success.

In related news, Water Canada recently published an article on the Great Lakes, highlighting the optimism of the Great Lakes and St. Lawrence Cities Initiative regarding the establishment of the CWA. The Canadian government has committed more than $650 million over the next decade to protect and restore water quality and ecosystem health in major watersheds across the country, including the Great Lakes. One transformative project under consideration by the CWA is called "For Our Waters." This initiative aims to produce satellite data for municipalities around Lake Ontario and validate the results to better identify sources of pollution. How reassuring is that?[174]

Some skeptics argue that satellite images are not required to determine industrial pollution, that the problem is evident where the pollution enters the water, and that resources would be better spent on upgrading water treatment plants to ensure all pollutants can be removed for effective water treatment systems for all Canadians.[175]

MDA Ltd., a leading provider of advanced technology and services to the global space industry, has been selected as the prime contractor for Globalstar Inc.'s new low earth orbit satellites. This contract, valued at $327 million USD (approximately $415 million CAD), includes the design, manufacture, assembly and testing of seventeen satellites, with options for additional satellites.

Critics argue that the $415 million spent on satellites could be better allocated to upgrade water treatment plants or on research to improve drinking water systems. Philippe Murphy-Rheume, Great Lakes and St. Lawrence Cities Initiative's Chief Development Officer, hopes that the "For Our Waters" project will help municipalities and conservation authorities target pollution sources. Some believe resources could instead be focused on improving analysis and effectiveness of water treatment processes to protect against PFAS and many other toxic substances, thereby identifying all communities that should be on DWAs.[176]

I like to share positives whenever I come across them. Lake Huron forests were not sprayed with toxic glyphosate starting this year (2025) thanks to high profile protests led by local First Nations, Stop the Spray Ontario and Traditional Ecological Knowledge Elders. The group set up a round dance blockade on Highway 17 at the Serpent River First Nation to demand the province stop the aerial spraying of glyphosate on their territories. This event, once again, reaffirms it is only when people demand that enough is enough, that we can expect positive change.[177]

OCEAN CLEANUP

At eighteen, Dutch inventor Boyan Slat founded The Ocean Cleanup to combat a critical environmental crisis: fourteen million tons of plastic entering oceans annually, threatening 230 million marine species.

Water advocacy extends beyond drinking water to protecting natural sources like oceans, which provide food for billions. Slat's nonprofit develops innovative technologies, including solar-powered trash-collecting barges, designed to remove oceanic plastic pollution. Their ambitious goal is to reduce worldwide ocean plastic by 90 percent, addressing the devastating impact on marine ecosystems.

His work demonstrates how young innovators can tackle global environmental challenges through technological solutions and determined action.

Chapter 49

Algae

In Lake Erie, the culprit of the algal blooms is phosphorous-rich agriculture and urban pollution—the bulk of which comes from the United States.[178] That combined with the shallow, warm water, creates the perfect habitat for the dangerous scum to thrive. The pollution persists despite Canada and the US striving to improve the water quality since the 1970s, and despite it being the source of drinking water for twelve million people.

Scientist Aaron Kirk is overseeing the project. "We take it for granted that we have all of this water," he said, "We need to be more serious about it. It is not an inexhaustible thing."

In the worst-case scenario, a severe toxic bloom could force a shutdown of water utilities. Even if it doesn't come to that, the blooms could disrupt the water treatment process, making it more costly and complex.

The analysis of many First Nations and rural community water treatment plants often overlooks their capability to remove all potential contaminants. Ensuring safe drinking water requires building effective treatment systems capable of handling any threats in the source water.

Left untreated, water with algal blooms can cause vomiting, nausea, diarrhea, pneumonia and fever. People swimming in such waters may experience skin rashes and blistering of the mouth. The need is obvious for continued efforts to address the Great Lakes' water quality challenges before the situation worsens.

Algal blooms can potentially affect many source waters and highlight the challenges faced by communities with inadequate water treatment systems. The most cost-effective way to ensure safe drinking water is to protect all source waters, making treatment systems less costly. This means banning or significantly reducing both agricultural and industrial runoff or discharge into waterways.

A massive outbreak of toxic algae off South Australia, which has devastated hundreds of species of marine life and disrupted local tourism and fishing, is a "natural disaster," stated Premier Peter Malinauskas.[179]

"I want to be really clear about this. This is a natural disaster," Malinauskas said. "I think politicians can do themselves a disservice when they get caught up in technicalities. This is a natural disaster. It should be acknowledged as such."

The algal bloom, first detected in March, spans an area 4,500 square kilometres in size and has been aggravated by rising ocean temperatures, environment officials say. More than 400 different species of marine life have been killed off or died as a result of the algal bloom, according to Malinauskas.

While algal blooms are naturally occurring, I don't believe they are natural disasters. Rising ocean temperatures and increased rainfall associated with climate change can exacerbate blooms, combined with runoff from agriculture and other human activities introduces excess nutrients into waterways, which can fuel rapid algal growth. We have ignored climate change for far too long despite scientists and activists begging for action from all levels of politicians.

NEVER GIVE UP

The First Nation of Na-Cho Nyak Dun (FNND) won its appeal against Yukon's approval of Metallic Minerals' exploration project in the Beaver River watershed. The Yukon Court of Appeal upheld the lower court's decision quashing the 2021 approval, ruling Yukon breached its duty to consult by refusing to engage on key issues: delaying project approval until the Beaver River Land Use Plan's completion and conducting community consultation.

The Court found Yukon failed to meaningfully respond to FNND's requests despite the deep consultation duty required. The decision sent the project back for reconsideration, establishing that decision-makers must justify refusals to accommodate Indigenous consultation requests meaningfully.

Chapter 50

Diquat

Around 1994, Dr. H_2O was asked to present to a small rural gathering of farmers. He was astonished to learn they were applying diquat to the dugouts on their farms as a means to control algal blooms. What bothered Dr. H_2O tremendously was that they also used their dugouts for drinking water! He shared his utter alarm with those present, at which point one middle-aged man broke down, emotionally distraught; he shared that his young son was in intensive care and doctors didn't know what was wrong with him. Dr. H_2O believed he was poisoned by diquat.

I don't recall what happened to the young boy or to the distressed farmer. I do distinctly recall that, just a matter of days later, Dr. H_2O was called in to the office of the president of the SRC, where he was bound and gagged, told never to talk to anyone in the public or the media about anything to do with diquat or Roundup ever again, or he would be fired. We had a mortgage to pay and a family to feed, so he did as he was told, or rather, as he was threatened.

That dictator-style attitude to science was one of the most blatant times that I am aware of, when science was abused rather than used for the greater good. That was when Dr. H_2O began planning to leave the SRC.

I always thought of SRC as being a scientific research company, but investment into innovation and technology commercialization remains its top priority.[180] Their budget includes $20.1 million to continue its work in "spurring economic growth" across a variety of industries in Saskatchewan including manufacturing, mining, agriculture, oil and gas, nuclear and critical minerals.[181] I used to assume that SRC's main objective was to further science. I was wrong, for decades. Their main purpose seems to be to aid in profiteering for corporations in Saskatchewan. And I am not the least surprised.

Dr. H_2O was determined to speak freely, and to warn those in vulnerable situations, including about products such as diquat or Roundup. He wasn't alone. Around the world, others were experiencing similar health effects and similar political denial.

When one side of his body seized up after working the fields of his small holding, Valdemar Postanovicz feared he was having a stroke.[182] "All the right side of my body was paralysed. I couldn't feel my foot and my hand. My mouth twisted to the right," he says.

In 2021, Postanovicz experienced acute pesticide poisoning after accidentally absorbing Reglone, a potent herbicide containing diquat, while clearing weeds from his land in a remote village in southern Brazil.[183] "It happened only once, but I felt so ill that I never used it again," he shared with Unearthed and Public Eye. Nowadays, he manually weeds his fields of beans and tobacco.

Postanovicz is among a growing number of farmers in Paraná, Brazil's agricultural heartland and largest consumer of the herbicide, who have been poisoned by diquat. Following the ban on the notorious weedkiller paraquat in Brazil in 2020, the country's usage of diquat, a chemically similar herbicide, has surged. Between 2019 and 2022, annual diquat sales in Brazil skyrocketed from approximately 1,400 to 24,000 tonnes, marking an increase of over 1,600 percent.

Reglone, a popular weedkiller brand in Brazil, contains 20 percent diquat and is manufactured by Syngenta in Huddersfield, England. Despite being banned on British and Swiss farms since 2020 due to high risks to nearby residents, British law still allows Syngenta to produce and export it to countries with weaker regulations.[184]

In Paraná, diquat usage has surged more sharply than in Brazil overall, leading to an increase in reported poisoning cases. From 2018 to 2021, the state recorded one to three cases annually, which jumped to six in 2022 and nine in 2023. Experts believe these numbers are just the tip of the iceberg, as many incidents go unreported due to lack of healthcare access in remote areas or fear of employer reprisals.

Marcelo de Souza Furtado, a specialist at the Paraná state health department, states that for each registered poisoning, there are likely fifty unreported cases according to the WHO. The true scale of the pesticide poisoning problem in Paraná remains unknown, but it is significant.

I am constantly amazed at how corporate greed turns a blind eye to the Brazilian farmer. Once again, it is the impoverished people, often people of colour, those who do not have the resources to adequately represent themselves, to fight for their own rights, who pay the dearest price. For me, this is a perfect example of government negligence and lack of due diligence.

More than two years after the European Food Safety Authority (EFSA) raised concerns about Syngenta's diquat, the Swiss agrichemical giant avoided an EU ban by challenging the watchdog's findings. Emails, letters and technical papers released by the European Commission show that the commission twice withdrew a proposal to ban diquat after Syngenta questioned EFSA's methodology.

Syngenta's lobbying in Brussels created a divide between the commission and its food safety agency, delaying the ban. Scientists and pesticide activists question why member countries and the commission have overridden EFSA's concerns. Medical officer Geoff Calvert suggests there are alternatives to diquat, and the decision to keep it on the market lacks legitimacy. Hans Muilerman from Pesticide Action Network Europe argues that reapproval is not possible if the hazardous components of diquat are considered.[185]

Syngenta claims it was simply pointing out weaknesses in EFSA's case, with spokesperson Anna Bakola stating that lobbying is a normal part of any political system.

Lobbying on behalf of company products should never, ever, be a normal part of any functioning political system, in my opinion. But *if* a product like diquat is going to be allowed, then the least any authorizing government body should do is demand that labeling clearly state the associated risks to educate and inform innocent members of the public to enable them to make informed decisions prior to using it.

Although diquat's license ran out in 2012, it has been extended in the EU on a temporary basis since then, in what Bakola describes as a "routine" process for such products.

An advertisement for a company selling diquat popped into my feed, as they do far too often—unwanted, unrequested, but sent to me by Facebook algorithms, the business expecting it to make them more profit. Alarm bells rang loud for me when I read their marketing ploy, "Become a Hero." The definition of a hero to a two-year-old is someone who buys them ice cream. A hero to an adult is someone who rescues a drowning person. A hero is a person who is admired or idealized for courage, outstanding achievements or noble qualities, such as "a war hero." Anyone applying diquat is definitely not a hero in my mind.

It is extremely hard to detect diquat in water, and you need to do a number of tests to find it.[186] Not all algae respond to specific chemicals in the same manner or time. Certain chemicals can inhibit or encourage algal growth. Murky waters, or waters with high dissolved organics, can also add to challenges applying chemicals and establishing their uptake. All of these factors contribute to concern over the application of diquat.[187]

I am not a scientist; I was the kid who knocked over the Bunsen burner in secondary school science. However, without understanding the scientific terms, it is abundantly clear that diquat pretty much kills everything in whatever water body it is used in. This will affect all the minuscule critters living in the water, each of them playing a critical role in the food chain and ultimately threatening wildlife. The scientific peer-reviewed papers demonstrating this were produced in 1999, more than twenty-five years ago. That the world hasn't funded further research to negate or dispel any corporate critiques of the research process, and that they continue to allow diquat to be used by some wannabe heroes, to me is criminal.

These advertisements are prepared by SePro, the company who operates as Stewards of the Water. I submitted a question to them: "I am interested to learn what chemicals you use in your treatment processes, and what qualifications people in your company have to provide the scientific treatments."

His prompt reply confirmed what my gut had suspected:

"Hi, Susan:

SePro has two products that are registered for use in Canada, Littora (diquat dibromide) and ProcellaCOR FX (florpyrauxifen-benzyl). SePro has been innovating, manufacturing, and selling aquatic herbicides and algaecides for thirty years. We have many on our staff with Ph.D.'s (Algae Scientist, Vascular plant Scientist, Formulation Chemists, etc.) Our Sales Team has hundreds of years of combined experience in developing and writing prescriptions for aquatic vegetation management. Thanks, and I hope this helps.

John Goidosik, Regional Manager—NorthWater Diagnostics & Restoration."

To which I sent another email: "I would like to know what concentrations you are applying, and if your workers are wearing PPE or are you advising anyone applying your product to wear PPE?"

I was completely open about who I was, my signature tag line states: Author of *Water Confidential, Witnessing Justice Denied—The Fight for Safe Drinking Water in Indigenous and Rural Communities in Canada.*

I was not surprised that he didn't reply.

I participated in their all-American webinar, touting the benefits of removing highly toxic algae blooms from source waters. I wasn't surprised that the chat was closed. They invited Q and A, but only two were submitted and neither were discussed. I asked if they were proud of removing the highly toxic blooms by using what some would say are highly toxic chemicals, and how can these toxic chemicals be removed in a water treatment process to protect drinking water.

Of course, they didn't answer. I didn't expect them to.

Another highly controlled webinar, it reminded me of the one I participated in for Water First. When an organization cannot be totally transparent, I believe they are hiding something.

I cannot fathom how trading toxic algae for toxic chemicals is an advantage, or that it makes anyone a hero.

Syngenta's campaign to keep diquat on the EU market took place largely out of public view, while a high profile political battle raged in the last couple of years across Europe over glyphosate—the active ingredient in a ubiquitous weedkiller made by American company Monsanto.

In a significant victory in the fight against toxins, environmental and health groups in Canada have successfully challenged the federal governments renewal of a glyphosate product, Mad Dog Plus. Friends of the Earth Canada, Safe Food Matters, the David Suzuki Foundation and Environmental Defence Canada, represented by Ecojustice lawyers, challenged Health Canada about the science used in assessing the "acceptable risk" of active ingredients in pesticides and to ensure it uses current research findings.[188]

Two Brazilian studies found that glyphosate, which occurred after glyphosate use increased due to genetically modified (GM) crops, transported through rivers was robustly associated with increased infant mortality and cancer deaths in children. Brazil uses almost twice the amount of glyphosate per hectare than the US. A US study stated: "The mounting evidence of the negative health externalities associated with the rollout of GM crops and the ensuing glyphosate intensification warrant new policy discussions about informed efficient and equitable regulation of these technologies."[183]

Can you see how removing contaminants in our drinking water treatment systems becomes a much greater challenge and, therefore, a much more costly process than if we restricted what chemicals affect our source water quality? Or how the cost of not protecting our source waters drastically increases our costs for health care? Or its impact on the way of life for all those affected by drinking unsafe water? And this is not including the environmental costs of the effects of polluting all life within our waterways.

STUDENTS TRANSFORM A FOOD DESERT INTO A "GREENBELT"

In Houston, Texas, East End, a food desert serving the district's poorest students, twenty high school and college students became "Green Ambassadors" over three years. They trained and certified through Project Learning Tree's environmental outdoor education, becoming among the first high schoolers to do so.

These young leaders aimed to provide over 100,000 residents with fresh, natural foods while creating habitats for wildlife and pollinators. They planted fruit trees and community gardens one at a time, linking schools and neighborhoods to form a Houston East End Greenbelt. In fall 2016, Furr High School won a $10 million grant through a national contest sponsored by Lauren Powell Jobs.

Chapter 51

Monsanto

Way back in 1998, I was driving on a gravel road toward Highway 219 early one morning, keeping a watchful eye for the deer that often ran out zig-zagging in front of me. My windows were wide open; I loved the fresh air on my face. Wide open farmland stretched on either side of me. CBC radio shared a story about a farmer named Percy Schmeiser from Saskatchewan in a battle with Monsanto over crops he was supposedly growing without paying for the seed.

Everyone living in the area had an opinion. Many took the side of Monsanto. Farmers buying Roundup Ready (RR) Monsanto canola seed were convinced they were growing food so the world wouldn't starve (actually, convinced isn't strong enough a word—they had been brainwashed to believe that they were the heroes.) They followed their religious convictions with the best of intentions, thankful for the gift of Monsanto canola. Where I lived, it seemed that one person in five believed Percy. In my remote location, there were only five homes, and I was the only one who believed him. I will always see Percy Schmeiser as a hero! There is a movie out now; Percy was released in theatres just days before his death on October 13, 2020. Percy's legal battle with Monsanto lasted for more than six years. The case began in 1998 and ended in 2004.

Monsanto had a patent on genetic modification to plants that rendered them resistant to glyphosate-based herbicides, such as Monsanto's Roundup. Branded as RR, these GM seeds allowed farmers to control weeds with a single pass of Roundup early in the growing season. The glyphosate would kill all plants except for the GM ones. Using a single herbicide generally decreases input costs, thereby increasing yields. RR seed was sold under a licensing program.

Farmers choosing to use it were obliged to purchase new seed every year. One of the terms of the licensing agreement was that farmers could not save part of the RR harvest and re-use it for planting a new crop in subsequent years. This was a departure from other conventional seeds that could be saved and replanted from one year to the next. But seeds from neighbouring farms had blown into Percy's field so that some of his crop was from seed that fell under the licensing program.

In Percy's own words at the end of his long legal battle: "I do not have to pay Monsanto one cent for profits, damages, penalties, court costs, or their technology user fee of $15/acre. I feel good about this ruling, as I have said all along that I didn't take advantage or profit from Monsanto's technology in my fields. I am pleased that the Supreme Court felt that way as well." Percy also stated: "Corporations should not be able to control people, seeds, plants, and food through patent law." I couldn't agree more.

I was fortunate to spend a month on the island of Molokai in Hawaii, soon after the collapse of the 2008 financial market. The condo I first stayed in had mostly locals living there, as the previous, mostly American, owners had been mortgage-rich and cash limited, thereby walking away from their vacation homes. The one-time resort now had an overgrown golf course, no maintenance and limited revenue. Prior to the resort, the locals had been gainfully employed by Dole, as one of the largest producers of pineapples. Now their social decline reminded me of my time spent in some First Nations communities back home. I noticed a sign plastered on a pole in the main part of town. Monsanto was inviting everyone to attend an information event, offering employment, security and wages that would feed and house their families again. As I took my place in line, the aura of hope surrounded me.

After their high pitch sales speeches, I stood up, asked for the microphone and introduced myself as coming from Saskatchewan, the home province of Percy Schmeiser. I expected Monsanto officials to act faster. I think I took them by surprise and was able to talk of their questionable product, and the potential harm to their water supply and to the food they grew and ate. The Hawaiians were not interested. All they saw was an opportunity for a better life, again.

And I couldn't blame them, but I knew I had tried.

A decade or more later, in 2016 and 2018, Monsanto agreed to have representatives in court to enter guilty pleas to the thirty-two environmental crimes related to using a pesticide on corn fields in Hawaii and storing a banned pesticide on Molokai and Maui. The company was given three years of probation and a fine of $12 million USD, including $6 million in fines and $6 million in community service payments. Monsanto was also required to continue an environmental compliance program with a third-party auditor for three years.

From the additional $6 million in community service payments, $1.5 million went to each of the following agencies: the Department of Agriculture for the Pesticide Use Revolving Fund—Pesticide Disposal Program and Pesticide Safety Training; the Department of the Attorney General for the Criminal Justice and Investigations Division; the Department of Health for the Environmental Management Division to support environmental health programs; and the Department of Land and Natural Resources for the Division of Aquatic Resources. US Attorney Tracy Wilkison said Monsanto repeatedly violated environmental laws, exposing people to harmful pesticides. EPA special agent Scot Adair said Monsanto's actions put people and the environment at risk. For example, between March 2013 and August 2014, Monsanto stored 160 pounds of Penncap-M hazardous waste at a facility on Molokai, despite knowing it needed disposal. Penncap-M is a restricted-use pesticide that cannot be purchased or used by the public and can only be used by certified applicators due to its potential adverse effects on the environment and risk of injury to applicators or bystanders.

Documents filed indicate that Monsanto was aware of the significant potential harm Penncap-M could cause to people and the environment. In addition to spraying the banned pesticide at one of its three facilities on Maui, Monsanto stored 111 gallons of Penncap-M at the Valley Farm, Maalaea and Piilani sites. This made Monsanto a "large quantity generator" of acute hazardous waste at these locations, according to court documents.

The company broke federal law by not using a proper shipping manifest to identify the hazardous material and by not getting a permit to accept hazardous waste at its Valley Farm site when it transported Penncap-M there in 2014. In a previous guilty plea and deferred prosecution agreement, Monsanto paid $10.2 million.

In 2019, Monsanto admitted to a misdemeanor for unlawfully spraying the banned pesticide Penncap-M on corn seed and research crops at its Valley Farm facility in Maui in 2014. According to the US Attorney's Office, Monsanto used Penncap-M despite knowing it was banned after 2013 due to an EPA cancellation order. The company also admitted that after the 2014 spraying, it instructed employees to re-enter the fields after seven days, even though they should have waited thirty-one days. Also in 2019, Monsanto pled guilty to two felony charges for unlawfully storing Penncap-M.

The company also plead guilty in 2020 to spraying a banned pesticide on research crops in Maui. In yet another case, Monsanto admitted to misdemeanor crimes involving a pesticide called Forfeit 280. In 2020, Monsanto allowed workers to enter fields in Hawaii sprayed with Forfeit 280 during a restricted period, violating their previous agreement.

Monsanto stated that it regrets its conduct and is taking steps to improve compliance, including stricter policies and enhanced training. The company said it is confident these measures will ensure legal compliance and maintain safety standards. If behaviour like this is maintaining high safety standards, I don't like to think what neglectful behaviour might look like.

Susan Blacklin

Fast forward to 2024, when the Mexican government amended its constitution to prohibit the planting of GM corn seeds. The reform enshrines the ban on GM corn planting in place since 2013, and mandates federal promotion of traditional agricultural practices through investment, research and institutional strengthening. The amendment comes on the heels of a December 2024 trade dispute panel ruling that Mexico's 2023 presidential decree, which sought to restrict the use of GM corn in dough and tortillas and phase out the use of glyphosate, violated the US-Mexico-Canada Agreement.[190]

In 2010, a US financial website revealed that the Bill and Melinda Gates Foundation had significantly increased its investment in Monsanto, purchasing 500,000 shares, worth around $23 million.[191] This news sparked outrage among critics, including Seattle-based AGRA Watch, a project of the Community Alliance for Global Justice. They criticized Monsanto for its history of neglecting the interests and well-being of small farmers globally and questioned the foundation's substantial funding for agricultural development in Africa.

The situation worsened when the African Centre for Biosafety, a watchdog based in South Africa, discovered that the Gates Foundation was collaborating with Cargill on a $10 million project to "develop the soya value chain" in Mozambique and other regions. This corporate jargon likely indicates the large-scale introduction of GM soya in southern Africa.

The two incidents raise many questions for the foundation. There are genuine concerns at both governmental and community levels that the extensive high-tech farming model used in the United States may not be suitable for most of Africa and should not be imposed on the poorest farmers under the guise of "feeding the world."

To be fair there are good things happening from GM organisms. Just like there are good things happening on the internet and with AI. I see that Monsanto's behaviour is to farming what the dark web is to the internet. Without government regulations or accountability, whether for agriculture, AI, care homes, the internet or for drinking water, I see the lack of regulations as problematic.

Did Bill Gates know his foundation was in danger of being caught

268

up in Monsanto's reputation or did the foundation actually share their corporate vision of farming and intend to work with them more in the future? The foundation has not been upfront about its vision for agriculture in the world's poorest countries, nor the role of controversial technologies like GM.

Percy was certainly a hero, but so too were all the individuals who pressured the Gates Foundation, so they sold their shares in Monsanto. One person can make a difference and, when many people stand up and make demands, they hold so much more power than any one person. We must fight for our rights, and also for the rights of those who are unable to fight for themselves.

But that is still not the end of the Monsanto saga. In a move considered to be retaliation for tariffs Canada placed on Chinese imports, including electric vehicles, in September 2024 China announced a dumping investigation into Canadian canola. "Dumping" refers to a trade practice where a country sells goods in a foreign country for a lower price than at home. Ian Boxall, president of the Agricultural Producers Association of Saskatchewan, mentioned that it's too early to predict if farmers will reduce their canola planting in the future, but some might reconsider seeding it. He added, "We'll see how volatile the market is next spring."

The European Union has created a legal framework to ensure that modern biotechnology, especially genetically modified organisms (GMOs), are developed safely. They aim to protect human and animal health and the environment by implementing a high-standard safety assessment at the EU level before any GMO is marketed. The EU has established harmonized procedures for the risk assessment and authorization of GMOs, ensuring these processes are efficient, time-limited and transparent. They also mandate clear labeling of GMOs to help consumers and professionals, such as farmers and food operators, make informed choices. Additionally, the EU ensures the traceability of GMOs on the market.

I am not surprised that the EU has taken this stance. They also have regulations for drinking water—protecting all water intended for drinking, cooking, food preparation or other domestic purposes—

whether in public or private premises. This includes water from any source, whether supplied through a distribution network, tanker or bottles, including spring waters. The regulations also cover all water used in food businesses for manufacturing, processing, preserving or marketing products for human consumption. Of course, the EU took this position; unlike Canada, they operate based on the precautionary principle.

To improve the safety of tap water in Europe, the EU is revising its Drinking Water Directive, which sets minimum quality standards for water meant for human consumption to protect against contaminants. In 2020, the European government supported an agreement to boost consumer confidence and encourage the use of tap water for drinking. The new law updates drinking water quality standards, including tightening the maximum limits for certain pollutants like lead and harmful bacteria. It also establishes minimum hygiene requirements for materials in contact with water, such as pipes and taps, to prevent contamination. Additionally, endocrine disruptors (chemicals that interfere with the body's hormonal system, potentially causing adverse health effects), pharmaceuticals and microplastics will be monitored through a watch list mechanism, allowing the EU to update surveillance based on the latest scientific developments. Only in the EU—pity! I hope that as PM Carney rubs shoulders with European leaders, perhaps some of their forward thinking will rub off on him and Canadians will benefit.

It is important to me that I have the right to choose what I consume. If nothing else, Canada must label all products that contain GMOs. Sadly, it is more likely to be impoverished, uninformed Canadians who must purchase the cheapest food and, therefore, often GM foods. If you can't follow an organic diet, another option is to choose to avoid GMOs. This involves cutting out grains like soy, corn and canola from your diet, which are commonly found in snack bars, cooking oils, sodas, margarine, cereals, baked goods and biscuits.

Many people do not think that GMOs are as bad as people make them out to be. They profess that if we can live better through scientific developments, then why not? I cannot judge. My logical brain questions how companies, such as Monsanto, can control the seeds to be benign. How do they ensure this trait isn't passed down the food chain, continued to people who eat the GM foods? I know far too many young couples faced with infertility and question whether there could be a connection. Apparently it's not just me who is making this connection, as reported by CBS news:[192]

A pesticide called chlormequat, linked to animal infertility, was found in 92 percent of non-organic oat-based foods tested, including Quaker Oats and Cheerios. The Environmental Working Group detected it in seventy-seven of ninety-six urine samples from 2017-2023, with levels rising recently. Studies show chlormequat can damage reproductive systems and disrupt fetal growth in animals, raising human health concerns. While EPA regulations permit its use only on ornamental plants in the US, the chemical is used as a growth regulator for cereal grains internationally, making crops easier to harvest.

I was surprised to find out that glyphosate is actually a patented antibiotic, which means it can destroy the soil's microbiome by affecting the shikimate pathway, a metabolic process used by bacteria. When this pathway is blocked, vegetables lose their ability to provide essential amino acids and alkaloids that our bodies need but cannot produce on their own. Glyphosate is water soluble, making it even more hazardous to our health as it can contaminate water systems.

Studies have shown that GMO foods contain fewer of these vital nutrients compared to their organic counterparts. This creates a double problem: our bodies can't get the necessary nutrients to detoxify and glyphosate residues also harm the gut microbiome, leading to inflammation and weakening the gut-brain barrier. This could explain the rise in autoimmune diseases and general health issues. Plus, these chemicals hinder the soil's ability to sequester carbon and contribute to our already unstable climate.

The most critical factor here for me, is that Roundup, Diquat and the myriad of chemicals now applied willy-nilly on crops all leach into our source waters. Runoff from productive fields is inevitable. While we would all love to go back to a time when kids drank water in creeks or from garden hoses, when we didn't worry about mercury in the fish we caught, the fact is that today our waterways are contaminated with diverse substances to various degrees. That the government cozies up to corporations and neglects the health and rights of its citizens, in my mind, is not acceptable.

SERIOUS ETHICAL CONCERNS

Science journal retracts study on safety of Monsanto's Roundup declaring, "Serious ethical concerns."

The journal *Regulatory Toxicology and Pharmacology* has formally retracted a sweeping scientific paper published in 2000 that became a key defense for Monsanto's claim that Roundup herbicide and its active ingredient glyphosate don't cause cancer.

Martin van den Berg, the journal's editor-in-chief, said in a note* accompanying the retraction that he had taken the step because of "serious ethical concerns regarding the independence and accountability of the authors of this article and the academic integrity of the carcinogenicity studies presented."

*https://www.sciencedirect.com/science/article/pii/S0273230099913715

Chapter 52

Nuclear Waste

It's not just pulp and paper, diamond mines or Big Oil that have a terrible impact on the health of Canadians. Nuclear waste also threatens all of us. A year after I emigrated to Winnipeg, I met a lovely couple who also came from the UK. They brought four children to enjoy new opportunities. Their eldest daughter recently reminded me how I once looked after them, how they loved to play dress-up around our apartment with my high-heeled shoes or play my vinyl records. Now I'm pleased to call Anne Lindsey a dear friend. She is also a passionate activist.

The rest of this chapter has been written by Anne Lindsey:

Shortly after the Liberals won the Canadian 2025 election, 130 civil society groups from across the country wrote to Prime Minister Mark Carney, reminding him that the "nation-building" energy and infrastructure that Canada needs will create good jobs and build the economy, and that it will also respect Indigenous rights and protect the climate. Unfortunately, our concerns seem to have fallen on deaf ears as the federal government plans to spend billions to speed up approvals of new (but old school) projects.

Building new pipelines and other oil and gas infrastructure to export fossil fuels, seems like a cruel joke as we choke on smoke from yet another "wildfire season" (a new name for a Canadian summer).

In Manitoba alone, more than 1 million hectares of forest have already burned as I write this in July of 2025—a disgraceful record facilitated by global warming-induced dryness across the region. Tens of thousands of mostly Indigenous people of northern communities have been displaced. This is not a joke, it's an emergency. And yet, more fossil fuels are on the menu, along with an alarming push for massive new investment in nuclear power from many parts of the country, including the federal government.

What's wrong with nuclear power development? After all, isn't it a "clean," "green," and "emissions-free" way to electrify our future? No. These are deceptions the nuclear industry would have us believe in their glossy propaganda campaigns being waged in print, digital and social media often targeted at young people expressing understandable anxiety over an uncertain future and searching desperately for alternatives.

Yet nuclear power is not an innovative alternative. Reactors take decades to build and go online. We do not have the luxury of time to avoid global tipping points of climate change. Nuclear is also the costliest way to generate electricity. Studies by organizations ranging from the Ontario Clean Air Alliance to Lazard—a global financial advisory and asset management firm founded in 1848—show that nuclear power is not competitive with renewable alternatives, which continue to drop in price. Every dollar invested in nuclear is a lost opportunity for developing cheaper and more readily available renewable energy, such as wind and solar combined with storage and smart grids.

The Intergovernmental Panel on Climate Change shows nuclear to be inefficient in reducing emissions and that it will only worsen our climate situation.

These are not the only problems with nuclear power. Both uranium mining and routine operations of nuclear plants produce a cocktail of daily radioactive emissions to air and water. These are experienced across the country, from northern Saskatchewan and Ontario's billions of tonnes of mine tailings, to the Great Lakes' burden of tritium from plants around their perimeter, to New Brunswick's Bay of Fundy, where the Point Lepreau reactor is located.

The fuel inventory of nuclear plants is highly susceptible in times of conflict (witness Zaporizhia in Ukraine). Despite the soothing moniker of the "peaceful atom," the entire nuclear fuel chain is inextricably linked to the proliferation and production of nuclear weapons, perhaps the ultimate threat to human existence.

But one aspect of the fuel chain that does not get a lot of mention is the waste products: the mine tailings and the end products—the highly toxic and long-lived, so-called "spent fuel" from the reactors and the reactors that become progressively more contaminated and eventually they, themselves, become radioactive waste at the end of their lives.

This is the topic that originally got me engaged in nuclear research, back in the 1980s, when the Canadian nuclear industry built an "underground research laboratory" here in my province of Manitoba. The purpose was to study the suitability of burying nuclear waste in the granite rock of the Canadian Shield, but local people were concerned about the possibility of the hole becoming a dumping ground for Canada's (and possibly international) nuclear waste.

Intensive lobbying of the provincial government at the time resulted in legislation banning the disposal of high-level radioactive waste in Manitoba. After all, Manitoba never actually saw benefits from nuclear electricity generation. Though a smaller-sized research reactor was built on federal land at Pinawa, it was shut down in the '80s and is still awaiting decommissioning. This is another deeply troubling story as the proposal is to essentially bury this radioactive building, which is basically on the shores of the Winnipeg River.

The URL was closed and the federal government went back to the drawing board about what to do with all the nuclear waste that continues to accumulate. They basically ended up with the same idea— a deep hole in the rock, in Northwest Ontario, ironically not far from the border with Manitoba. And this is where the story picks up—some forty years later.

An earlier version of what follows was written by Anne Lindsey and published in the Spring 2025 edition of The Cottager *magazine, based in Winnipeg, and is reprinted here with permission:*

One of the undeniable joys of being at the cottage, the cabin, camp, or simply "the lake," is the water itself. Whether paddling, fishing, swimming, or just savouring the reflected sunset, we all experience that deep appreciation and connection to the water—nowhere more than in Northwestern Ontario and Southeast Manitoba. Folks who live there year-round, including those who are Indigenous to it, undoubtedly experience that connection in an even deeper way. Water is life-giving, a source of sustenance and livelihood, intimately bound up with the cycles of the land, and a spiritual home.

So, when a decision is made that may threaten the water, it's something we need to pay attention to. Something that we want to be consulted about. Such is the case with the November 28, 2024, decision to cite a Deep Geological Repository (DGR) in Revell Township, situated between Ignace and Dryden, in Northwest Ontario. It's meant to store virtually all of Canada's high-level nuclear waste—the deadly radioactive fuel from nuclear reactors. About 130,000 tonnes of it is currently in temporary storage at reactor sites, most in southern Ontario, New Brunswick and Quebec, thousands of kilometres from Sunset Country.

What is the connection to water? The Revell site lies at the headwaters of the English-Wabigoon and the Turtle-Rainy River systems. Both flow into the Winnipeg River (the latter through Lake of the Woods) and then into Lake Winnipeg. Does it make sense to store radioactive waste in the Lake Winnipeg watershed? More about that later. First, why this site?

The Nuclear Waste Management Organization (NWMO), an

industrial consortium of nuclear waste owners in Canada, has been seeking expressions of interest for hosting a burial site for over a decade. The industry believes that placing the waste hundreds of metres underground in a stable rock formation is the best and safest way to isolate this toxic material from the biosphere for the hundreds of thousands of years that it remains dangerous. This hypothesis has never been proven (waste production only began in the mid-20th century for the development of nuclear weapons) but it's the approach being taken by several countries (such as Sweden and Finland) who have their own deep burial research underway. Meanwhile waste has been accumulating at reactor sites around the world, both in water-filled pools—needed to reduce its extreme heat—and in concrete storage containers.

Ignace, Ontario was one community that put up its hand, some of its residents seeing this as an economic development opportunity. NWMO's process stipulated that siting would only proceed with a "declaration of willingness" from a host community. After years of discussions, funding of infrastructure and financial contributions to the town, and a hotly contested "public consultation" process. Ignace held a vote and declared itself a willing host in October.

A multimillion-dollar signing bonus was exchanged for some dubious promises, including an agreement to accept any future nuclear waste and never to disagree publicly with NWMO. The other community that NWMO consulted was the Wabigoon Lake Ojibway Nation (WLON) whose traditional territory includes the Revell area. It is not known publicly what incentives were offered to WLON during the consultations, but a vote of both on- and off-reserve WLON citizens in November resulted in consent for continued study of the area (something the NWMO refers to "as site characterization.") It's important to note that WLON's news release states that the yes vote "does not signify approval of the project ..."

This is where NWMO's site selection process went off the rails—while they had not, in fact, received declarations of willingness from both communities, they announced a siting decision anyway. And despite years of study and assessment that lie ahead, Canada's nuclear industry began to loudly celebrate the "solution" to the waste problem.

Even though many other communities stand to be impacted, only Ignace and WLON were formally consulted.

The waste must be transported from southern Canada, requiring three massive truck shipments per day for the next forty years, for the existing waste alone. The route skirting the east and north shores of Lake Superior, through Thunder Bay and westward is known to be treacherous, especially in winter. Small municipal or volunteer emergency crews are often first responders to accidents, yet no communities along the route were consulted. Many have publicly stated firm opposition to the risk. And what about communities downstream from Revell? That's where our water connection comes in.

WLON is in the Lake Winnipeg watershed, Ignace is not.

A massive surface transfer facility will be key to the DGR complex. There, highly radioactive, thermally hot waste will be removed from transport containers and transferred to burial canisters. The now-radioactive transport containers will be "cleaned" and returned for another load. Any radioactive releases from routine operations or potential accidents could compromise surface water flowing downstream. Many of the isotopes involved are water soluble —some can substitute themselves for their non-radioactive counterparts found in plants and in animal (including human) bodies, becoming a permanent threat to the food chain and drinking water.

Even if surface incidents miraculously never occur during the decades-long lifespan of the facility, there is still a major concern about groundwater. Deep rock formations are not as dry and solid as we may imagine. Hard rock miners know that water flows deep underground through cracks and fissures, albeit slowly. To maintain adequate space between the hot canisters, massive excavations are required. The disturbance may exaggerate cracks, allowing faster flow of groundwater. The industry acknowledges that it's a matter of time before the canisters are compromised and radioactivity enters groundwater. This threatens recharge zones of aquifers and surface waters and all the downstream ecologies and communities that depend on them. Small wonder that thirteen downstream Treaty 3 First Nations have declared their opposition, even though they too were never formally consulted.

Industry scoffs at these concerns, saying there is "no watershed" deep underground. Yet why the massive investment of billions of dollars and years of research, be that on canisters, backfill material or the rock itself, all focused on keeping radioactivity out of water, if they can reassure us today that deep burial is safe now, and will be for thousands of years? Nobody can predict the future. Modelling can be deceptive.

And now it seems clear that Canada is set to build more reactors, including a plan for the country's largest nuclear facility in Peace River, Alberta, as of fall 2025 this project has been put "on hold," but it's still in the cards...also, the Federal government have just announced $2B for the new small modular reactors (SMRs) built at Darlington, through Bill C-5.[193] Other newer projects still in the design stage, many of which use novel fuels, and all of which will add to the ever-growing stockpile of waste in temporary storage at reactor sites. This puts the lie to the industry's pious claims that underground "disposal" is morally necessary to get the waste off the surface, where it's currently stored—because evidently there will *always* be waste in storage at the surface—and just recently the NWMO announced it's now looking for another "willing host" community for yet another repository. Will this qualify as a "nationally important project"? Will it also be in Northwest Ontario, or in someone else's watershed?

The social and ethical question for all of us—cottagers, year-round residents, taxpayers, Indigenous Nations—is do we want to accept this profound risk to the future of water on behalf of those who come after us? - *Winnipegger Anne Lindsey has been anti-nuclear for nearly forty years.*[194]

A few more words from me, Susan Blacklin now, on this issue. In 2024, Kebaowek First Nation brought a case before the Federal Court in Ottawa against Canadian Nuclear Laboratories (CNL), contesting the approval of a Near Surface Disposal Facility.[195] This facility, intended for construction on Kebaowek's traditional territory, would house over 1 million cubic tonnes of low-level nuclear waste just a kilometre from the Kichi Sibi (Ottawa River).[196]

The proposed project poses serious risks to culturally significant species and ancient forests, and it threatens to pollute a watershed that supplies drinking water to more than 9 million people. Moreover, the facility was approved without obtaining Kebaowek's Free, Prior and Informed Consent, as mandated by Canada's UNDRIP Act.

In February, the court ruled that UNDRIP should have been respected when the Canadian Nuclear Safety Commission authorized CNL's nuclear waste facility on Kebaowek's land. This positive decision reaffirmed that governments must seek the Free, Prior and Informed Consent of Indigenous communities. CNL quickly appealed the ruling, prompting Kebaowek to file a cross-appeal in response.

Perhaps, those in power will soon pay attention. I suspect that Kebaowek First Nation will soon spearhead new precedents for consulting with First Nations. Kebaowek First Nation faced off against Canada's premier nuclear organization in October 2025 at the Federal Court of Appeal in Ottawa, in what's shaping up to be a big case with far-reaching implications for industry and project proponents under federal jurisdiction.[197]

Anne's persistence for forty years is amazing but not surprising! It has been almost thirty years since we founded the SDWF. Maybe in another ten years some politicians will begin listening.

Chapter 53

Canada's Ongoing Commitment to Nuclear Energy

I hope this gives readers an indication of the huge commitment the present governments are making to nuclear energy. Governments in Canada, at the federal, provincial and territorial levels, continue to dedicate significant financial and other resources to the further development of nuclear energy in Canada, both for large-scale nuclear and SMRs. At the federal level, some recent examples include:[198]

- $304 million in loans to finance the development and modernization of a new, large-scale CANDU nuclear reactor and support the broader Canadian supply chain;
- $970 million in investments to date from the Canada Infrastructure Bank to advance the Darlington SMR Project;
- $69.9 million for Natural Resources Canada to support activities to minimize waste generated from SMRs, support the creation of a fuel supply chain, strengthen international nuclear cooperation agreements and enhance domestic safety and security policies and practices;
- $50 million in federal funding from the Electricity Predevelopment Program to support the Bruce Power's

assessment of new generation opportunities contemplated for Bruce C;

- $250 million, over four years starting in 2022–23, for predevelopment activities of clean electricity projects of national significance, such as SMRs;
- $50.7 million for the Canadian Nuclear Safety Commission to build its capacity to regulate SMRs and to work internationally on regulatory harmonization; and
- Up to $80 million in funding to Saskatchewan's Crown Investments Corporation to SaskPower's SMR predevelopment work.

GRAND SALMON SOURCE TO SEA

In July 2022, Libby Tobey, Hailey Thompson, Alia Payne, and Brooke Hess reached the ocean after paddling the length of the Salmon River while promoting the removal of the four dams and a moratorium on the Stibnite Gold Project to save Idaho's rapidly dwindling salmon populations.

Along the way, they paddled 1,000 miles, organized community members to write thousands of letters to Congress to protect the river, and raffled off one coveted date with pro kayaker Benny Marr.

Chapter 54

Salmon

Scientists have proven the problems for decades. Citizens have protested for decades, but our government listens to lobbyists and corporations rather than to its scientists and citizens. Here is another example of the unacceptable length of time it took the government to react.

Alexandra Morton is an American and Canadian marine biologist best known for her thirty-year study of wild killer whales in the Broughton Archipelago in British Columbia. Since the 1990s, her work has shifted toward the study of the impact of salmon farming on Canadian wild salmon. I listened in awe of her strength, inspiration and resilience as she spoke at an International Women's Day event.

Alexandra Morton came north from California in the early 1980s, following her first love: the northern resident orca. In remote Echo Bay, in the Broughton Archipelago, she found the perfect place to settle into all she had ever dreamed of, a lifetime of observing and learning what these big-brained mammals are saying to each other. She was lucky enough to get there just in time to witness a place of true natural abundance, and she learned how to thrive in the wilderness as a scientist and a single mother.

Then, in 1989, industrial aquaculture moved into the region, chasing the whales away. Her fisherman neighbours asked her if she would write letters on their behalf to the government explaining the damage the farms were doing to the fisheries, and one thing led to another. Soon, Alex had shifted her scientific focus to documenting the infectious diseases and parasites that pour from the ocean farm pens of Atlantic salmon into the migration routes of wild Pacific salmon, and then to prove their disastrous impact on wild salmon and the entire ecosystem of the coast.

Alex stood against the farms, first representing her community, then alone and, at last, as part of an uprising that built around her as ancient Indigenous governance resisted a province and a country that wouldn't obey their own court rulings. She has used her science, many acts of protest and the legal system in her unrelenting efforts to save wild salmon and, ultimately, the whales.

It is a story that not only reveals her own doggedness and bravery but also shines a bright light on the ways other humans doggedly resist the truth. She brilliantly calls those humans to account for the sake of us all.

The Canadian federal government has announced that it will ban open-net salmon farms in British Columbia starting in 2029. This plan includes renewing over sixty licences across the province for another five years and making closed-containment salmon farming systems eligible for a nine-year licence extension. This announcement follows the federal government's commitment five years ago to phase out open-net pen Atlantic salmon farms.

It has been decided that facilities will be expected to make restocking and harvesting decisions that are consistent with the requirement to "fully terminate" all open-net pen farming by June 30, 2029.

What I don't understand is how a government committed to making closed-containment salmon farming systems eligible for a nine-year licence extension.

It is shocking that it took forty-five years before the government reacted to Alex Morton's sound research to protect the wild salmon and begin closing fish farms. Forty-five years! The song keeps playing over and over in my mind, when will they ever learn, when will they ever learn.

David Attenborough recently released what could be his final documentary, titled *Ocean with David Attenborough*. He specifically chose to use the word "Ocean" instead of the word "oceans" in the title to emphasize the interconnectedness of the world's waters and the life that relies on them.[199]

It's long been known that deep seabed mining is bad news for ocean life. Two new studies: "Race to the Bottom" and "Mining for Trouble" estimate that while annual royalties distributed among member countries of the UN Convention on the Law of the Sea would range from US$42,000 to US$7.35 million, mining economies would lose more than US$560 billion annually, because deep sea minerals cannot be taxed.[200]

I suggest it is similar for surface and groundwater: we only have one water, what happens in place A affects the water in place B. His statement is also very compelling, "After living for nearly one hundred years on this planet, I now understand the most important place on earth is not on land … but at sea."

Similarly, I believe that protecting our water is the top priority for protecting human health, the environment and climate change.

WAKING THE PUBLIC

Amid outcry over sewage discharged along England's coastline, grassroots campaigners scored notable inland victories against water companies. Ash Smith, founder of Windrush Against Sewage Pollution (WASP) in Oxfordshire, UK, stated rational conversation with authorities failed, but people power, hard evidence, and media exposure drove change.

In June, WASP's groundbreaking investigative work helped environmental nonprofit Wild Justice take water regulator Ofwat to court over alleged failure to stop raw sewage discharges into rivers.

The case's crowdfunding campaign reached its £40,000 target in one day. Hard-working groups like WASP awakened the public to England's deteriorating rivers and seas, leading to an explosion in concerned citizen activism and accountability efforts.

Chapter 55

Thames Water

I imagine many of you are familiar with the River Thames running through London, England, where I grew up. When Thames Water was privatized in 1989, it had no debt. But over the years it borrowed heavily and its total debt (which includes all of its borrowings and liabilities) now stands at GBP 22.8 billion, according to the latest financial results.

The executives, shareholders and private equity companies who own it, have presided over decades of under-investment, aggressive cost cutting and huge dividend payments.

The result of these decades of mismanagement can be seen in the scale of sewage discharges, the record leaks from its pipes, and the state of its treatment plants, all of which are now at the centre of a criminal investigation and a regulatory inquiry by Ofwat.

Ofwat is the Water Services Regulation Authority, the economic regulator for the water and sewage sectors in England and Wales.[201] It ensures that water companies provide good quality and efficient services to consumers at a fair price. Ofwat also works to ensure the long-term resilience of water supply and wastewater systems and to promote competition in the market. I can't help but wonder if Ofwat employees have been snoozing or playing cribbage instead of monitoring Thames Water over the years.

Britain's biggest water company was fined almost 123 million pounds (USD $166 million) in May 2025 for releasing sewage into rivers and streams while paying dividends to its shareholders.[202]

Consumers and politicians have criticized the company, arguing that Thames Water created its own problems by paying overly generous dividends to investors and high salaries to executives while failing to invest in pipelines, pumps and reservoirs.[203] Company executives say that the fault lies with regulators, which kept bills too low for too long, starving the company of vital cash to fund improvements.

Privatization was intended to lead to improved water quality and lower bills, but instead it turned water into a cash cow for investment firms and private equity companies.[204] Thames Water has now accumulated insurmountable debt, while issuing dividends to its shareholders like candy.

Thames Water has now been fined over GBP 120 billion, equivalent of $165 billion, mostly for releasing at least seventy-two billion litres of sewage into the River Thames since 2020. This is roughly equivalent to 29,000 Olympic-sized swimming pools of crap. There are a few more people living in London, England than in Winnipeg.

Here in Canada, over 200 residents in a rural area northwest of Penticton, British Columbia, may have to shell out over $1,000 a month just to have water.[205] The jaw-dropping cost has homeowners that rely on the Sage Mesa Water System stressed out and demanding a more reasonable solution. Last year, the 242 residents of Sage Mesa were notified that they would be on the hook for the $33 million cost of upgrading the waterline. The system does not meet basic water treatment guidelines set out by Interior Health and is reaching the end of its life with much of the infrastructure over sixty years old.

While Sage is in a similar dilemma to London in the UK, I see similarities between citizens in both places. The working people are paying outrageous prices for the basic sustenance of life, because of dividends and/or profits being shared with predominantly wealthy investors. Though not officially privatization, it is not operating for the greater good of the people. Though somewhat hidden and not so blatant, it is still sucking the cash out of the infrastructure designed to deliver safe drinking water to Canadians.

Individuals in Canada are benefitting hugely from these financial disbursements, similar to the UK dividends, while communities' drinking water treatment systems continue to fail. An army of volunteers have stepped up in the UK, under the organization named Windrush Against Sewage Pollution. *Take Back our Water* is their slogan.[206]

While water treatment in Canada may not have been officially privatized, too many are making extreme profits while the shameful situation of unsafe drinking water prevails, as most systems in Indigenous communities and in many in other communities across Canada remain ineffective.

We may have water in abundance, but we aren't immune to freshwater shortages and Canadians are some of the world's highest domestic water users.[207] Metro Vancouver's average water consumption is 270 litres per day (LPD), per person, which is higher than the Canadians average of 223 LPD. Yet by the 2050s the western mountains' anticipated snowpack depths, a significant drinking water source for Metro Vancouver, is projected to decrease by 60 percent!

I suggest that here in Canada we too must "Take Back Our Water," to ensure all Canadians have access to truly safe drinking water and protect our source waters.

Chapter 56

Sewage

Between 2013 and 2023, Winnipeg discharged 115 billion litres of sewage into its river system, which is equivalent to filling nearly 46,000 Olympic swimming pools.[208] That is approximately 1.5 times (in ten years) the 72 billion litres (in five years) of sewage discharged by Thames Water in London England between 2020 and 2025. Thames water treatment serves 16 million residents, Winnipeg serves 936,000 residents. Winnipeg dumps an average of 10 billion litres of sewage into the rivers every year, compared to an average of 12.2 billion litres annually in the Thames![209]

With climate change expected to bring more intense rainstorms, flash floods and temperature fluctuations to prairie cities like Winnipeg, these overflows are likely to become more frequent. Lake Winnipeg, the final destination of the Red River's journey from the Minnesota-North Dakota border, is the sixth-largest lake in Canada and the eleventh largest in the world. In 2013, it was named the "most threatened lake of the year."

Algal blooms have closed beaches and popular fishing spots as the water turns green and is hazardous to touch. The lake's once-thriving fish populations have been strained by multiple challenges, leading to the closure of several fisheries. Dave Taylor, a passionate kayaker and concerned Winnipeg resident, wrote in a 1992 letter to the editor of the *Winnipeg Free Press*: "Our rivers are putrid and contain ten pathogenic organisms responsible for diseases such as salmonella, dysentery, ear infections, dermatitis, respiratory infections, gastroenteritis, and polio." [210]

Between 2013 and 2023, Winnipeg's seventy-six sewer outfalls overflowed an average of 1,300 times each year. Each outfall overflowed about fifteen times annually, releasing nearly 8 million litres of diluted wastewater—a mix of runoff and sewage—into the river system, totalling over 10 billion litres each year. This does not include an additional fourteen million litres released annually due to accidents and equipment failures.

Water sampling on the Red River and the Assiniboine River shows that the concentrations of suspended solids, phosphorus and *E. coli* almost always exceed provincial standards after a sewer overflow. *E. coli* levels can rise above three million units per 100 mL, while the provincial guideline during an overflow is 1,000 per 100 mL and the federal guideline for safe swimming is 235 units per 100 mL. I ask you, what good do provincial guidelines or provincial standards for water quality do to protect Canadian citizens?

Phosphorous is considered the main cause of algal blooms, and the North End sewage treatment plant is the largest point source of Lake Winnipeg's phosphorus overload—hence the urgency to tackle billions of dollars' worth of outstanding upgrades. Sewer overflows, by contrast, contribute very little to the overall phosphorus picture on the lake. "This is going to sound funny to say because we hear these horribly large amounts of raw sewage being discharged into our rivers … but daily, all the sewage that we send through the treatment plant dwarfs what's being released from the combined sewer overflows," says Alexis Kanu, the executive director of the Lake Winnipeg Foundation.

Winnipeg's mayor and council have taken the position that the rivers are big enough to dilute the toxins in sewage overflows, and murky and fast-moving enough to discourage swimming, waterskiing, and other water sports, so the risks are minimal—at least if you weigh up the costs.[211]

Water sampling on the Red and Assiniboine rivers shows the concentrations of suspended solids, phosphorus and *E. coli* nearly always exceed provincial standards after a sewer overflow. *E. coli* levels in particular can rise above three million units per 100 mL. The provincial guideline during an overflow is 1,000 per 100 mL. The federal guideline for safe swimming is 235 units per 100 mL

According to the foundation's analysis, sewer overflows are responsible for less than 0.5 percent of the lakes' total phosphorus load, while the treatment plants account for about five percent. Still, according to Dimple Roy, water management director at the International Institute for Sustainable Development, "The phosphorus pulses are not trivial. Cumulatively, we're still putting way too much phosphorus into Lake Winnipeg."

"The other things we would be concerned about are things like pharmaceuticals, pathogens like *E. coli* and then just a lot of organic matter," she adds. "The more we have these spills, the more we see our downstream waterways compromised."

In November 2023, the city realized one of two old pipes running under the river near the Fort Garry Bridge in the St. Vital neighbourhood was at risk of failing, so it shut the pipe down for repairs and diverted the sewage flow into its equally outdated twin. The second pipe failed three months later. More than 230 million litres of raw sewage flowed into the Red River over a period longer than two weeks, marking the second-worst spill in city history.

In response, eleven First Nations communities downstream of the river and surrounding Lake Winnipeg filed lawsuits worth a combined $5.5 billion against all three levels of government, alleging they have breached Treaty and Charter rights by failing to address Winnipeg's decades of water pollution. The three suits, originally filed separately, are now being litigated together.

"Treating the Red River and Assiniboine River as part of the sewage system has polluted Lake Winnipeg," the statement of claim says.

The case is incredibly complex, encompassing discharges from the treatment plants, sewer overflows and emergencies like the 2024 sewage spill, over a period of about twenty years. The nations say repeated sewage releases into the river—and the impacts on Lake Winnipeg—have caused health problems, destroyed fisheries, limited access to drinking water, prevented traditional practices and had adverse psychological effects, including a mistrust of the waters.[212] The communities allege they have not been informed when leaks or accidental releases occur, nor has there been meaningful dialogue about the way these leaks and spills impact the First Nations. The eleven nations are bringing the case forward as trustees of the lake, known as Weenipagamiksaguygun, who "has a spirit as a living being."

"The city has a history of repeatedly violating its environmental licences and having wastewater discharges that are purportedly accidental or unplanned but entirely predictable," the claim alleges. "Every discharge of treated or untreated wastewater into the river system, whether planned or unplanned, constitutes a continuous and ongoing breach of duty by the city."

To build some resilience in the urban landscape, stop the flow of harmful pathogens into the city's waterways and make the rivers something residents can proudly enjoy—the city will need to move much faster than planned.

"We should be celebrating our rivers here in Winnipeg. We should be using them and enjoying them," Kanu says. "They should be a source of joy for us—not a source of fear and contamination."

I suggest that Winnipeg is not alone, that many other cities, maybe as many as to say *all* other cities, and no doubt many towns, are also faced with an abundance of sewage and failing infrastructures. If you can't swim in the water, if you are afraid of becoming ill, then too much pollution has impacted the health of that water.

This sign was posted near Goderich, by a public beach on Lake Huron. I ask you, does everyone know not to swallow the beach water? There was a time when kids could play in the water in Lake Huron or in Lake Winnipeg, but not anymore!

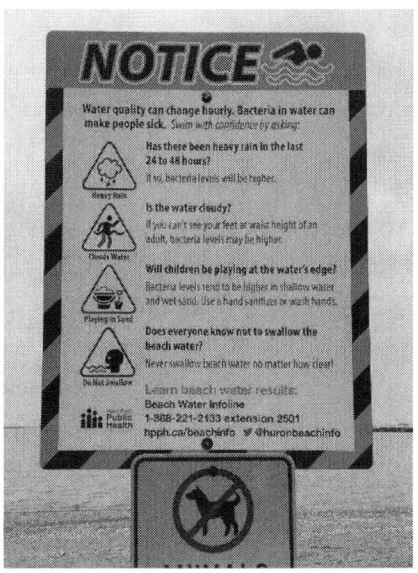

In the summer of 2025, residents were being told to avoid certain shorelines in Saanich and Oak Bay, British Columbia, because of a health risk posed by the presence of raw sewage. Oak Bay, British Columbia, has now obtained over $8.8 million from federal, provincial and local governments to replace the nearly one-hundred-year-old, single-pipe sewer system in the Uplands neighbourhood, which currently combines sewage and stormwater that can overflow into the ocean.[213]

The single-pipe system collects sewage and stormwater together, but heavy rains cause excess water that treatment plants cannot handle, leading to untreated sewage overflowing into nearby water bodies. High levels of *E. coli* and fecal matter frequently cause beach and water body closures. Not only in Oak Bay but in communities across Canada, with over 167 billion litres of combined sewage and stormwater discharged into water bodies nationwide in 2017.

I wonder if that includes the ever increasing cruise ship industry. I'm not convinced they only empty their tanks when they dock in Canadian cities. But then private boats and yachts also dump out at sea.

A groundbreaking project to prevent diluted sewage from

contaminating local waterways by eliminating combined sewer overflows into the Don River will transform this watershed's health.

Toronto has launched the largest and most ambitious stormwater management program in the city's history. With a budget exceeding $3 billion, this initiative will dramatically improve water quality in the Lower Don River, Taylor-Massey Creek, and Toronto's Inner Harbour. The program upgrades technology and expands capacity to capture, transport, and treat combined sewer overflow before it reaches our waterways. Construction is already well underway.

Kingston, Hamilton, Ottawa, and Sarnia have also committed to multi-year plans and programs addressing combined sewer overflow. This work is slow and expensive, progressing as budgets and resources allow, but it represents a critical investment in protecting public health and restoring the ecological integrity of our urban waterways.[214]

Chapter 57

Rural Representation

Indigenous people are represented by the AFN. Who is standing up for rural Canadians? I wonder where provincial organizations representing rural communities across Canada position themselves with regard to safe drinking water. Silence. Seeing as British Columbia leads the way of all provinces with DWAs surely the UBCM should be taking action on their part. I see their conference in Victoria September 2025 opened with Marcy Grossman as their keynote speaker.

Her speech topic: Charting the Course. "Transforming her career from criminal psychologist to Canadian diplomat, she also shares that 'impossible is possible - if you have the right mindset and the right tools.'" Of course it is! I am sure Marcy is a super qualified psychologist and diplomat/ambassador. Achieving safe drinking water for all Canadians isn't impossible, it may just seem that way to people wondering where all the government funding has gone. Maybe her criminal psychology background is more appropriate to determine the answers.

To make safe drinking water possible, all we need is political will and due diligence by those responsible. We need elected politicians to walk their talk, we need accountability. I wonder if her keynote encouraged the UBCM to act on behalf of the communities they represent. We need politicians and organizations like UBCM to initiate change. Perhaps they do and government ignores them too, after all, they ignored their resolution in 2019 calling for a halt to water extraction licenses. In the absence of those qualities, we need Canadians to stand up and demand better.

Safe drinking water is probably as big of a problem, or maybe even a bigger problem, in rural communities as it is in Indigenous ones. I expected that the Federation of Canadian Municipalities (FCM) would be concerned about drinking water in towns like Princeton. A board member of FCM explained she couldn't bring my book *Water Confidential*, or our petition, to FCM's attention as it is against policy for a board member to do so, to avoid anyone furthering their own agenda.

She suggested I ask our local municipality to put forward information; they ghosted my request. I tried submitting a request for contact through the FCM online portal, it kicked back as undeliverable. I tried phoning; a man said he would pass on my message. I received no response. They appear to have galas, dinners and speeches in common with SFNWA and WF, and like those charities do nothing to advocate or protect source waters or strive for safe drinking water for their members. The board member promised to read information I sent her. At the time of going to print, you guessed it, silence! As far as I can see, FCM is a bureaucratic behemoth that is modelled on our government.

There are many reasons people can't stand up and object, including lack of resources, busy lives, lack of information, lack of interest, hopelessness and feeling overwhelmed or insignificant. Add in other issues, such as food insecurity, unemployment, disabilities and drug addictions, and it's a wonder anyone can fight anything anymore.

First Nations are particularly overburdened with assaults on their communities from many directions. How do they choose which assault to fight first? This is why it is so important for those who can object, protest and organize petitions to do more than their share. We must speak for those who cannot and hold those without morals and values accountable.

Chapter 58

Asbestos Cement Water Pipes

Statistics Canada data releases seldom attract much attention. June 24, 2025, was no different. "Canada's Core Public Infrastructure Survey: Required renewal budgets, 2022" yawned the headline of the Statistics Canada feature in *The Daily*. It was a deep dive into all things infrastructure in Canada. Roads, bridges, water mains, etc. Outside of infrastructure nerds, it was of interest to very few others. A line near the very bottom of the release on that day is important.

"In addition, estimates of the length of asbestos cement pipes that carry drinking water are available upon request." The survey represents the first time Canada has bothered to tally up its asbestos cement water pipes. The pipes, which were installed in Canada from the 1940s into the 1990s, were once popular because of cost and light weight. They contain approximately 20 percent asbestos.

The Statistics Canada data shows that as of December 31, 2022, there were 13.7 thousand kilometres of aging asbestos-cement pipes delivering water to millions of Canadians, with British Columbia having the most at 4,261 km. The information was compiled by Prevent Cancer Now (PCN)—a Canadian NGO which operates entirely with volunteers—after it filed a parliamentary petition calling for the question on asbestos cement water pipes to be included in a national survey. The disregard for regulating asbestos cement pipes and the effort by petitioners raises questions about other avenues for Canadians to dispute federal negligence. Results such as this make the effectiveness of another petition to demand national drinking water regulations uncertain, given the influence of lobbyists and potential corruption.

The tale of how we got here is both fascinating, disturbing and an excellent example of political denial, misinformation and obfuscation. It involves an individual by the name of Julian Branch. In 2012, the Regina resident had discovered Regina has 600 kilometres of old asbestos cement water pipes. Because he is trained as a journalist and professional communicator, and he sits on the board of PCN, Julian started asking questions. A lot of questions. This chapter follows his queries and his findings.

Some background to explain how we got here is helpful. In 1973, a landmark environmental court case was under way in the United States. Asbestos measuring 644 million fibres per litre (644 MFL) had been discovered in Lake Superior, near Duluth, MN. The Reserve Mining Company had been dumping iron ore tailings into the lake for decades, causing the problem. It was ordered to stop the practice by the fledgling EPA. While the trial was ongoing, environmental groups discovered that America had installed a lot of asbestos cement water distribution pipes. A two-decade long investigation was launched into asbestos in American water.

In 1992, the EPA regulated asbestos in water. Just five years earlier, the Department and Health and Human Services had released a comprehensive study entitled "Report on Cancer Risks Associated with the Ingestion of Asbestos." The conclusion begins with an almost dismissive tone. "We conclude that sufficient evidence is not available for a credible quantitative cancer risk assessment of asbestos ingestion at this time." However, a few sentences later the *conclusion* takes on a decidedly different tone. "Even if the increased rate of cancer is less than ten percent of the background rate and cannot be demonstrated by available research tools, the ingestion of water, food or drugs laden with asbestos by millions of people over their lifetimes could result in a substantial number of cancers. Several members of this working group believe it is prudent, preventative health policy to recommend eliminating possible sources of ingestion exposure to asbestos whenever and to whatever extent possible," reads the document. It adds that one approach to solving the problem would be "eliminating asbestos cement pipe in water supply systems."

The 1992 American regulation set the Maximum Contaminant Level Goal at seven MFL. A 1992 EPA fact sheet says the level was established "to protect against cancer." A 1995 EPA fact sheet says the long-term health effect of swallowing asbestos is "lung disease: cancer."

Canada hadn't just been watching. It had launched its own studies in the 1970s. Chief among them was "Asbestos and Drinking Water in Canada," which was published in 1981. It showed there were asbestos fibres in Canadian waters, but more importantly it talked about the "Contribution of asbestos cement pipe." The report from the Department of Health and Welfare Canada highlights Winnipeg, Manitoba. It says the concentration of asbestos in Winnipeg water was as high as 6.5 MFL. The report goes on to say: "This data provides a statistically valid indication that erosion of asbestos cement pipes is taking place." The federal government knew by the late 1970s that asbestos cement pipes were deteriorating and shedding fibres into the water.

In the 1980s, a report entitled City of Winnipeg "Technical Advice on Corrosion of Asbestos Cement Pipe" was commissioned. It found that the concentration of asbestos in Winnipeg water had almost doubled to 12.3 MFL. The authors of the report recommended testing for asbestos at ten locations on an annual basis. In a 2021 email, the City of Winnipeg confirmed that there are more than 700 kilometres of old asbestos cement water pipes in the city and that the city had stopped testing the water for asbestos in 1995. In an April 4, 2022, the City of Winnipeg committee meeting the director of water and waste confirmed that the highest concentration of asbestos in water prior to the cessation of testing in 1995 was nineteen MFL.

However, he was quick to point out that that was okay because the fibres were of the shorter variety. Winnipeg has not tested its water for asbestos for three decades.

Regina has 600 kilometres of old asbestos cement water pipes. The city describes the condition of the pipes as "sound." City information also shows there were 2,477 asbestos cement water pipe breaks in Regina between 2010 and 2022. The city maintains it has not detected any asbestos in Regina water despite testing since 2016.

CTV's investigative television news program *W5* tested Regina's water near one large main break in 2022 and sent the sample to an American lab. The results showed there were 370,000 asbestos fibres in a litre of water. When the program aired in March 2023, journalists across Canada fanned out to learn what the situation was in their communities.

A television news broadcast from Vancouver Island told viewers there were hundreds of kilometres of old asbestos cement water pipe in Nanaimo, British Columbia. The news story informed viewers that the city tested the water for asbestos, while the proud mayor took a big swig of water on-camera, further assuring residents that it was safe to drink.

FOI documents show that Nanaimo tested the water for asbestos, at one location in 2007. City documents show that asbestos cement water pipes have continued to break since then. Mayor Leonard Krog, a lawyer and former provincial NDP cabinet minister, would have been aware of all of this when he said that Nanaimo water is tested annually for asbestos.

There is misinformation, disinformation and a complete and utter lack of information on this file. West Vancouver falls into the latter category. A homeowner had learned second-hand that his old asbestos cement water pipe was being replaced. He had heard about Julian's work in this area and reached out for assistance. He wrote a letter to the District of West Vancouver asking six simple questions, such as: as a homeowner, why wasn't he informed that the pipe being replaced at his home was made of asbestos cement, and could the district please provide the location of other asbestos cement water pipes in his community? The response from the Director of Engineering and Transportation Services with the District of West Vancouver is breathtaking: "Regarding the specific questions in your previous email attached one and three through five, please be aware that due to limited staff resources, requests for detailed information that is not readily available will not be addressed if it requires staff time to research, analyze and compile. While we understand this is not the answer you had hoped for, we thank you for your understanding."

What the author is saying essentially is that if your query requires any effort at all, we simply won't answer it. Done.

West Vancouver is home to some of Canada's priciest real estate. Do some of those properties have old asbestos cement water pipes? With 7.5 kilometres of the pipe in the neighbourhood, the answer is yes, but we have no idea which ones.

As mentioned, the homeowner's correspondence contained six questions. The director of engineering didn't even reference number six in her response. The final question referred to federal government studies, which show old asbestos pipes shed fibres into the water, and that those fibres can cause cancer. The homeowner wanted to know if the district intended to follow-up on those studies.

The Canadian Water and Wastewater Association (CWWA) represents Canadian water utilities. As Executive Director Robert Haller explains, the CWWA is "The voice of the utility sector to the federal government." On March 22, 2023, just two days before *W5* aired its segment on asbestos cement water pipes, the CWWA released a document entitled "Speaking Notes on Asbestos Cement Water Pipes." The four-page information paper contains the statement that asbestos cement water pipes were "used for a brief period in Canada during the 1940s through to the 1970s." It goes on to say that asbestos cement pipe hasn't been installed in municipal systems for over fifty years. The city of Regina installed asbestos cement water pipe until

1988. Winnipeg, Manitoba installed the pipe into the 1990s. These facts add decades to the CWWA statement.

The CWWA says asbestos fibres may be released into drinking water from runoff of mining tailings, improperly disposed contaminated household wastes and stream and groundwater contact with asbestos-bearing bedrock. It's clear the language is lifted directly from the Health Canada website. However, Health Canada includes one possible source which CWWA neglected to mention in its speaking notes.

"Asbestos fibres may also be released from asbestos cement pipes that carry drinking water from the treatment plant into your home," reads a Health Canada infographic related to asbestos in drinking water. The CWWA document also assures the reader that "Standard water treatment can effectively remove these asbestos fibres from drinking water supplies." This is little solace to the homeowner with a deteriorating or broken asbestos cement water pipe metres from their dwelling.

The 2023 CWWA document is illustrative of the misinformation surrounding this issue. Because the CWWA speaks to the federal government on behalf of Canadian utilities it needs to do so in an honest, truthful, transparent and inclusive manner.

In 1989, Health Canada had decided that there was no need to regulate asbestos in water. Its official position is "no evidence of adverse health effects from exposure through drinking water." Because of this, few municipalities measure asbestos in drinking water. The lack of a standard regulatory test method means Canadian labs may produce inaccurate results.

May 22, 2003, is when this started to get serious. Saskatchewan's top political and academic leaders had gathered at the University of Regina to launch a new $30 million NRC facility. It was to be overseen by something futuristic sounding called Communities of Tomorrow. The skillfully written three-page news release makes no mention of asbestos cement water pipes. Other than vague references to "municipal water management best practices," and helping the City of Regina "meet its existing and future infrastructure challenges," there is no sense of exactly what this new Centre for Sustainable Infrastructure Research (CSIR) would be doing. The federal government contributed $15 million toward the new centre, the Saskatchewan government, the City of Regina and the University of Regina each ponied up $5 million.

CSIR released its first study in June 2005. "Failure Conditions of Asbestos Cement Water Mains in Regina" left little doubt what the focus of the new research centre was. The twelve-page study says asbestos cement water pipes in the Saskatchewan capital "are experiencing more and more failures in recent years and account for almost all of the water main breaks in the city." The top of page two refers to asbestos fibres in water as a "health concern."

One month later, the University of Regina magazine UResearch wrote a glowing feature on CSIR. It said the 535 kilometres of "pressurized water pipes" under Regina's streets are experiencing ever increasing levels of failure. (It's worth noting that NRC documents from the same period show that Regina had exactly 535 kilometres of asbestos cement water pipes.) The University of Regina magazine article goes on to say that the facility will be permanent and that a team of twenty-five to thirty researchers will be working hard to solve a problem impacting a big chunk of the western world. "The same challenges exist in urban environments all over North America. The problem can only grow, making cost-effective assessment of buried infrastructure critical to municipalities ability to sustain our standard of living," cautions the article.

In 2008 and 2009, then minister responsible for the NRC, Jim Prentice, wrote a very positive summary of CSIR, saying that it was focused on water infrastructure, including performance of water mains, and that it planned to "assess the long-term performance of Asbestos Cement Pipe in North America and provide a guidance document for sustainable management of these assets in partnership with nineteen municipalities in North America." The federal government invested another $4 million in CSIR in 2009, bringing the funding total to $34 million.

The year 2010 was a busy one for CSIR, which produced a number of studies on asbestos cement water pipes. One of those studies warned of health concerns associated with "the inhalation of airborne asbestos fibres from showers, humidifiers, etc." As a matter of fact, the words "health concerns" are used in at least three of the NRC studies when referring to asbestos fibres in water. Yet another 2010 NRC study stated, "Severely deteriorated AC pipes also released fibers into the drinking water and could pose a hazard of malignant tumors of the gastrointestinal tract and other organs in consumers." (Note: while some sources use the acronym "AC" for asbestos cement, this book does not.)

The year was best summed up by scientists working at that research facility when they issued a one-pager calling for "immediate" and "urgent" steps to be taken in order to "ensure a safe and clean drinking water supply" in Regina. Despite the warning, according to City of Regina documents, the length of asbestos cement water pipe in Regina remained the exact same until at least 2021. When work is performed on asbestos cement water pipes in Regina today, residents receive an information sheet which reads: "When the water service is restored, the water may be discoloured as repairs stir up sediment from the bottom of the main. This disturbed sediment does not pose a health concern and should not make you sick." The city is right about one thing: drinking water should not make you sick.

When Julian discovered this issue in 2012, he took his concerns to politicians and the media. An early story came from *Global News* on May 30. It contains the line: "The report says minor disturbances like ground shifting can release some of the fibres into the drinking water. While inhaling asbestos is a major health concern, microbiologist Roy Cullimore, worked on that report, says there is no evidence ingesting it is harmful." Dr. Cullimore had co-authored two of those NRC studies in 2010 and 2011, which said asbestos fibres, from old water pipes, posed a health concern and could cause cancer.

On June 7, 2012, the City of Regina issued a media advisory for a hastily called scrum at city hall. "City of Regina and Ministry of Environment remind residents that city water is safe despite asbestos cement pipes," blares the headline. The media scrambled down to hear municipal and provincial officials assure them that despite the fact that Regina did not test the water for asbestos, that it was safe to drink. The headline of the story in the Regina edition of the short-lived *Metro* newspaper the next day was "Officials insist Regina water is safe." It includes quotes from the then-executive director of the municipal branch of the Saskatchewan Ministry of the Environment, Sam Ferris.

"There has been some talk about a study coming out of the United States that has looked at a linkage between asbestos in drinking water and some benign forms of cancer of the stomach. We don't know how rigorous that study is and we really need the time to look at that and any other emerging information." At the time he made those comments, Mr. Ferris had been the Saskatchewan representative on the federal/provincial/territorial drinking water panel for years. He would have been aware that there were several American reports linking the ingestion of asbestos to cancer, and that they had been written decades before he made his comments. One would think that would provide ample time to test their rigor.

PCN and the Canadian Environmental Law Association filed an Environmental Petition questioning Health Canada's stance on asbestos in water. Health Canada replied that the studies were not relevant to health because they focused on infrastructure.

The so-called permanent CSIR facility quietly closed its doors in 2013. All the research material and reports associated with the Communities of Tomorrow project were to be archived on the Saskatchewan Urban Municipalities Association's (SUMA's) website, according to a partnership deal agreed to by the two entities. However, when contacted by CTV News in 2024, the municipal organization said in a written statement "SUMA has no records of any reports done by Communities of Tomorrow related to asbestos cement water pipes, nor any mandate to do so."

The NRC studies were produced in Regina between 2005 to 2012. They contain a massive amount of data on asbestos cement water pipes. When Julian wrote the NRC to ask some elementary questions about asbestos cement water pipes in 2021, He received a surprising response. "NRC does not hold any data on asbestos cement water pipes," reads the October 4, 2021, letter.

At this juncture it will be helpful to briefly explain Canada's long, complex relationship with what used to be called the "miracle mineral." The information is found in a 1977 report from the now defunct Science Council of Canada. *Regulatory processes and jurisdictional issues in the regulation of hazardous products in Canada* contains some important background of asbestos in the country where it was once mined and exported. It says that Canada once produced 40 percent of the world's chrysotile asbestos, the most popular type used in the pipes, and that the province of Québec accounted for about 80 percent of that. However, more importantly it mentions the regulation of asbestos. "Among the six cases discussed in this report and the science council's policies and poisons study, the asbestos case is perhaps most indicative of the inadequacies of the Canadian regulatory processes, both at the regulation-making and the compliance levels," reads the report.

In the years leading up to a much-anticipated asbestos ban, the federal Canadian government promised to "ban asbestos and asbestos-containing products by 2018." However, a stakeholder from the cement pipe industry lobbied the federal Liberal government. As a result, the regulations do not apply to products containing asbestos already in use before the regulations came into force. Aging asbestos cement pipes in the ground under Canadian towns and cities were exempt. On October 18, 2018, Environment Minister Catherine McKenna said, "None of these exemptions will impact on human health—that is our top priority."

As of this writing, Health Canada was conducting another review of that 1989 decision not to regulate. The September 2022 guidelines for Canadian Drinking Water Quality state that a common source of asbestos in water is the "decay of asbestos cement pipes," but a guideline value is not necessary because there is "no evidence of adverse health effects from exposure through drinking water."

There is growing evidence that asbestos cement water pipes have reached the end of their useful lives, and they are experiencing an increase in breaks and failures around the world. A story out of Kenya from September 2025 said the replacement of old asbestos cement water pipes was being prioritized because "Scientists believe asbestos used in roofing and piping may be linked to gastrointestinal cancers in humans."

The article goes on to quote Cabinet Secretary for Water, Sanitation and Irrigation Eric Mugaa. "The Aguthi Water Project was launched in the 1970s using asbestos cement pipes, which are now very old. Asbestos pipes are believed to be responsible for some of the health problems we are facing, including cancer, and therefore we want to replace them," he told the media.

The asbestos cement pipe process was the invention of Dr. A. Mazza of Genoa, Italy in the early 1900s. Italy has a lot of old asbestos cement water pipes. Earthquakes in recent decades have wreaked havoc with underground water infrastructure. Some of the most recent research in this area is coming from Italy.

The 2016 study "Possible Health Risks From Asbestos in Drinking Water" concludes; "Several findings suggest that health risks from asbestos could not exclusively derive from inhalation of fibres. Health hazards might also be present after ingestion, mainly after daily ingestion of drinking water for long periods."

Another 2022 Italian study entitled "Better Safe than Sorry? Stated preferences and the precautionary principle for securing drinking water quality in an Italian district" contains a paragraph, which strikes at the very heart of this issue. The study points out that asbestos is not identified as a contaminant in drinking water in either Italy or Europe. "Nonetheless, public concern regarding asbestos in general is high, and the issue has attracted the attention of NGOs and political exponents. This is a case in which some public distrust and suspicion towards experts is understandable, given the legacy of hesitance in acknowledging the polluting potential of asbestos in the past, and the tremendous delay in taking action, which resulted in countless victims."

A recent investigation into asbestos cement water pipes and ingested asbestos by a consortium of journalists in the UK and Europe shone a light on this issue. "The types of cancer that you get from ingestion would be gastrointestinal tract cancers. Oesophagus, stomach, even small bowel, and then large bowel cancers, which increasingly are being shown to be related to asbestos, again, that data goes back to the 1960s and early 1970s, said Dr. Arthur Frank, professor of public health and professor of medicine at Drexel University in Philadelphia, PA, in a 2023 interview with EUobserver.

"Scientists, in general, had been [more] concerned about the inhalation of asbestos, which is how most people are exposed. And little attention, and little review, and little research has been funded that looks at the role of ingested asbestos," continued Frank, who is widely regarded as one of the leading international experts in this area.

All this takes us back to that innocuous Statistics Canada release from June 24, 2025. There are at least 13.7 thousand kilometres of old asbestos cement pipes delivering water to millions of unsuspecting Canadians, particularly in older neighbourhoods that may house more vulnerable, lower-income populations. These old pipes were exempt from a ban on asbestos in Canada.

The lack of general knowledge about the ingestion of asbestos can be attributed to politicians and government officials who fear the prospect of replacing all asbestos cement water pipes at a huge cost. Despite decades of studying the health hazards of ingested asbestos, little has been done to educate the public or physicians about these risks. Pipes made of asbestos cement that carry water to millions of homes can be deadly, according to numerous studies and reports.

"This issue will only get worse, as asbestos cement water pipes continue to age, deteriorate and break," says Julian Branch. "The issue of misinformation and disinformation attached to this file is sobering. Canada once led the world in research into asbestos cement water pipes. It's time we stopped denying that drinking asbestos is a problem and started to pay more attention to the research we funded two decades ago. If political leaders are looking for a true nation-building pipeline project, they need look no further than asbestos cement water pipe replacement."

The American Cancer Society states that when people drink water that flows through asbestos cement pipes they are at increased risk. The issue of asbestos in water is a significant concern, and the lack of regulation and testing raises questions about the safety of drinking water in Canada. The need for urgent action to address this issue is evident and the government must take steps to ensure the safety of drinking water for all Canadians. National drinking water regulations which include protecting source waters would be a good place to start.

Chapter 59

Asbestos Cement Pipes—To Remove or Not to Remove

If those you elected to look after your community choose to wait until asbestos fibres are found in your water samples, cancers may already be changing your life. Generally speaking, asbestos cement pipes are in the older neighbourhoods, which are quite likely now impoverished neighbourhoods. It will be interesting to see how residents of West Vancouver react, where the old neighbourhood is now home to some of the wealthiest people in British Columbia. You can check if your community has reported having asbestos cement pipes delivering your drinking water from the spreadsheet posted on PCN's website at preventcancernow.ca/where-are-canadas-asbestos-cement-drinking-water-pipes-check-the-data-for-where-you-live and on my website: susanblacklin.com.

Although First Nations communities fall under federal jurisdiction, Stats Canada decided not to include them in this project. Just one more example of how colonial control over Indigenous people is alive and well in the federal government, and shameful neglect of an opportunity to participate in true reconciliation.

Nick Garland is a Fellow of the Faculty of Asbestos Assessment and Management, the Founder of Assure 360 and Managing Director of Assure Risk Management Ltd. Nick shares his vast knowledge on this incredibly important topic. This is how Nick clarifies his concerns:

"Ingested asbestos has long been dismissed in the UK as posing no risk, with 100 percent focus on the inhalation hazard. I am very much a product of this thinking. Professor Frank's talk was very convincing. He highlighted the wide range of academic reports that detail an asbestos link to a wide range of cancers (ovarian, peritoneal mesothelioma, kidney and various gastrointestinal tract cancers). He also discussed the evidence of migration of fibres. Either the fibres are migrating from the lungs—or they are migrating from a combination of the lungs and the digestive tract. The latter seems much more logical to me.

"While I accept that the data is not 100 percent conclusive—the claim that ingested asbestos represents *no* risk is no longer supportable. How much risk? As yet I do not know, and I am not certain how it can be quantified as disentangling it from inhaled exposure is likely to be very difficult. But as the exposure is likely to be widespread, even a tiny risk is likely to have a significant outcome. Consequently, I believe that we need to start testing drinking water (especially in soft water and acidic ground water areas). I also believe that this will inevitably lead to a regulatory demand to replace all the affected pipes within a stated timeframe. Leaving pipes in the ground in the UK will not be allowed as that would change the location to an unlicensed waste site—not to mention the effect on land value. Removal to a special asbestos waste site as part of the replacement project will be demanded. Or ideally permanent destruction as is being researched and developed in the Netherlands."[215]

Nick continued: "This is not through any better regulations on asbestos cement water infrastructure—as I say this has to date been ignored or dismissed—but more the general requirements of UK law on what to do with asbestos waste once you have created it."

Decommissioned asbestos cement pipes which are aged or failing, where the binding materials have leached, can be up to 80 percent asbestos. When pipes are decommissioned and the water that ran through them is removed, the asbestos-containing pipes collapse, fracture and degrade into small pieces, becoming friable asbestos.

Friable asbestos poses a risk of asbestos exposure to workers. When disturbed, fibres can be released into the air and inhaled. Accumulated exposure is a well-documented risk factor for several diseases, including asbestosis, lung cancer and mesothelioma.

Both municipalities and contractors need to consider the management of decommissioning asbestos-containing pipes in the work zone. Section 6.7 (1) of the Occupational Health and Safety Regulation requires friable asbestos-containing material in the workplace to be controlled by removal, enclosure or encapsulation.

Re-burying asbestos-containing pipes in soil does not effectively encapsulate or enclose the asbestos, so you must ensure that degrading pipes are removed from the work area and disposed of appropriately. It is also mandatory to keep an exposure control plan for asbestos that includes safe work procedures for removal of asbestos-containing pipes.

A 2010 NRC study warns against this practice. "Abandoning the pipe sections or burying them in place is not recommended practice," reads a line from the Safety and Waste Management of Asbestos Cement Pipes article.

"This practice creates potential ACM [asbestos-containing material] waste exposure zones in the urban environment."

Removing asbestos-containing pipes is the most effective way to mitigate the risk of workers' exposure during current and future projects. Safe disposal methods must be used when removing uncovered decommissioned asbestos-containing pipes. Full PPE must be worn by workers and the area cordoned off to avoid accidental inhalation by the general public.

Future excavations will encounter the old pipes, creating a risk of exposing workers to airborne asbestos fibres if the pipes are not properly removed and disposed of. Improperly abandoned pipes can contribute to soil and groundwater contamination as they continue to degrade over time. Leaving these pipes in the ground may save money today but may mean that workers and the environment pay a much higher price in the future.

We have been aware of asbestos harmful effects for many years. "Asbestos causes cancer; it is a lethal carcinogen."[216] Consuming water from a leaking asbestos cement water pipe is dangerous. Regardless of how asbestos is ingested, exposure is hazardous. Further studies on the negative effects of asbestos on human health and the environment are not required. What is required is an accurate assessment of the state that asbestos pipes are in currently and an acknowledgment of the health hazards they pose to residents and workers.[217]

We need to take immediate action to address the issue of underrepresented research in asbestos cement water pipes and stop downplaying the threat they pose. We must follow the science provided and stand at the forefront of this topic on a national level to gain federal funding, as we are the ones with something to lose in this situation. Our health and the health of our children depend on it!

Illness due to asbestos can take decades to surface and it is easy for this government to make you believe it has no effect on your health. The repercussions of not acting now, knowing the dangers, will be catastrophic; the odds of many people fatally falling ill due to this exposure are extremely high and widespread.

A half century ago, the man widely regarded as one of the foremost experts on asbestos issued a similar warning during that landmark American environmental court battle involving asbestos in water. In 1973, Dr. Irving Selikoff told the world that ingesting asbestos was as bad as inhaling it and to do nothing would be to play "a form of Russian roulette, and we don't know where the bullet is," he said back then. "If we're wrong the consequences would be disastrous." Society chose to shoot the messenger instead of heeding his science. Not much has changed in fifty plus years.[218]

What Dr. Selikoff said fifty years ago holds even more truth today. Here in Canada asbestos removal and any legal requirements are defined by each province. I encourage readers to pressure their community leaders to do the right thing.

I find it interesting that asbestos companies filed for protection under a legal process known as bankruptcy reorganization, protecting them from getting sued in court. To receive this protection, the asbestos companies had to put aside billions of dollars to pay current and future claims. These are the funds that pay people diagnosed with mesothelioma. I wonder why oil companies and companies like the Dryden Paper Mill have not been required to establish contingency funds to provide effective water treatment systems to those they affected.

If you are concerned about your community not removing asbestos cement pipes appropriately, check your provincial laws. In British Columbia, the Occupational Health and Safety Regulation, Section 6.7 (1) states all asbestos cement pipes must be removed not buried. WorkSafe British Columbia (WSBC) then enforces this law. However, WSBC only responds to complaints, so don't hesitate to contact your provincial equivalent organisations. Furthermore, WSBC legislates that the prime contractor, that is usually the municipality or town, are responsible for specifying removal not burial for all asbestos cement pipes.

If a reader is aware of any community in Canada who are doing the right thing and removing failing asbestos cement water pipes, please let me know, they deserve to be honoured for their ethical values.

Part Six

Advocacy

Chapter 60

The Importance of Advocacy

Advocacy is more necessary today than ever before. As federal politicians begin to discuss how to open provincial borders to maximise trade between provinces, this is the perfect time to recognize that regardless of where people live, *all* Canadians have a right to the same high quality of drinking water. For those who have declared previously, "I don't do politics," you need to rethink your position. If you are not part of the solution, you remain part of the problem. If you think the corporations are too big, too strong, too powerful, let me tell you a story.

I visited a petting zoo with my grandchildren. My six-year-old granddaughter was keen to learn to read and asked endless questions. Her hands were carefully holding the food pellets she caught from the machine after we inserted twenty-five cents, intent not to drop even one pellet.

"Nana, what does that sign say?"

"Don't feed the baby goats."

"Who put the sign there?"

"Probably the people who own the baby goats."

"Why can't I feed them?"

"They probably have special food for baby goats."

A little boy around her age wandered up to the pen with the baby goats. He had spiked hair, a black mock leather jacket, and cowboy boots. Both children watched the baby goats intently; the boy began to reach out to feed the baby goats. My granddaughter moved toward him, declaring, "The sign says you don't feed the baby goats."

He stared back at her as if he is saying: *And who are you to tell me what to do?* A tall muscular man, over six feet tall and 200 pounds, wearing a sleeveless tank top to display many tattoos, sauntered up to the boy.

"What's up, son."

"She says I can't feed the baby goats."

"Them don't look like baby goats to me," said the man, as if encouraging his son to go ahead and feed them. My granddaughter tipped her head so far back she could have been studying the moon. He was so tall, but she made eye contact with the man.

"The sign says don't feed the baby goats, so the owner says they are baby goats, then they are baby goats and we don't feed the baby goats." The man and his son slowly walked away. I am sure they said she was bossy.

I was so proud of her. This wasn't about feeding the baby goats; it was about her standing up for her rights and for the rights of the baby goats who couldn't speak for themselves. I hope she never has to assert such confidence in inappropriate situations, but whenever she is in a tough situation, I love that she has practice in how to stand up for herself—and for those unable to stand up for themselves.

I love that some of my genes may have been passed down to her. If you are unsure about standing up, asserting your rights and protecting those less vulnerable, please think of my six-year-old granddaughter. If she can protest, so can you.

How we protest is the key. How can we be most effective? After great consideration, over the last six months, and many serious and contemplative discussions among our core group organizing this petition, under today's biased, self-serving bureaucratic undemocratic system we are questioning the best way forward. How can we best be heard?

If you, the reader, are outraged, and I think you should be, then call your MP and your MLA, not just today as this is fresh in your mind, but daily, weekly or monthly. Call and write to make them as sick of hearing from you as we are sick at the state of drinking water and ultimately of our democracy in this beautiful country. That is the only way we are going to possibly affect change.

One person can affect change, but the more people who join forces, the more people who contact their MPs, MLAs and municipal leaders. The stronger we are, the sooner we can be heard and demand what is our right: safe drinking water and source water protection in a fair and just society.

Globally, billionaire wealth grew by $2.8 trillion in 2024 alone, equivalent to roughly $7.9 billion a day, which has risen three times faster in 2024 than 2023. Meanwhile, the number of people living in poverty has barely changed since 1990. In 2024, billionaire wealth increased by $309 million *per day* while 3.8 million people live below the poverty line in Canada. Canadian billionaire wealth could easily carpet most of Vancouver (93 percent) in Canadian $50 notes.[219]

It is time to drastically defund many government agencies; the dilemma is whether they will continue funding celebratory dinners and galas or focus on improving the quality of drinking water for Canadians. Implementing laws for source water protection while increasing taxes for the extreme wealthy would be a good start. Canada's total national debt on May 29, 2025, is $1,261,698,486,073. That is one trillion two hundred sixty-one billion six hundred ninety-eight million four hundred eighty-six thousand seventy-three Canadian dollars. And too many Canadians still do not have access to safe drinking water. Fires are raging and climate change is real. We must advocate for responsible, accountable, governance now more than ever.

Chapter 61

E-Petitions

A call for a no-confidence vote against Justin Trudeau and the Liberal government has become the most signed e-petition in Canada, gathering 300,000 signatures and highlighting the growing popularity of online petitions.[220] Since their introduction in 2015, the number of signatures on e-petitions presented in the House of Commons has steadily increased, reaching nearly 1.5 million names on about 500 petitions in 2024.

In December 2025, a Change.org petition to Stop Silencing Survivors: Ban National Security Secrecy in Violence Against Women Cases, should not have been necessary to protest, receiving over 9,000 signatures in a short time. Neither should it be necessary for federal ePetition #7027 protecting Civil and human rights. It received over 28,000 signatures before it closed. Will our government heed or act on either petition? On any petition?

Although e-petitions can draw attention to public issues and request action, they do not legally obligate the government to change its policies. The government must provide written responses to petitions presented in the House, usually outlining current policies or laws.

Only Canadian citizens can sign e-petitions. Given their growing popularity, many believe it is time to review the rules to make them harder for governments to ignore. Kennedy Stewart, a former NDP MP and Vancouver mayor, who initiated the motion for e-petitions, suggested that if a petition reaches a certain number of votes, it should automatically trigger a Commons "take-note" debate or be addressed by a special committee. He noted that e-petitions are very popular in the UK, where the House of Commons has a committee to handle them. Stewart emphasized that with the decline in democracy worldwide, e-petitions are a good way to strengthen our core values.[221] I would like to see e-petitions follow the same principals of referendums as in Switzerland and empower Canadians to demand change.[222]

Jonathan Cassels from Kitchener, Ontario, created one of the most popular e-petitions, "Implementing Electoral Reform," which surpassed the second most signed, e-petition by over 40,000 signatures. Launched in November 2016, it gathered 130,452 signatures and was sponsored by NDP MP Nathan Cullen.[223]

Cassels asked the government to:

1. Commit to making the 2015 election the last under the First Past the Post system.

2. Outline reform proposals.

3. Consult the public on these proposals.

4. Provide a timeline for introducing an electoral reform bill.

The petition received significant support from residents of Ontario, British Columbia, Alberta and Québec. Although electoral reform was a key promise in the 2015 campaign, the government abandoned the idea in February 2017, citing a lack of broad support. The government's response echoed this sentiment and stated that changing the electoral system was no longer in the minister's mandate.

If changing the electoral system was not in the Minister's mandate, then what business did Trudeau have in running his election platform on this promise, effectively buying votes? Furthermore, I would like to see the survey results for which Minister Karina Gould told reporters it had "become evident that the broad support needed among Canadians for a change of this magnitude does not exist." I question if such a survey exists?

Another big part of Trudeau's platform was ending all long-term DWAs in First Nations communities. Neither of those promises came to fruition. Then again, do any politician's promises come to fruition?

Did you know that a Canadian prime minister with a majority of the seats—a majority government—has more control and power than a US president or a UK prime minister? Strange but true. The US Constitution created separation of powers, separating legislative and executive branches. In Canada, a prime minister with a majority government holds virtually all the power, controlling both the legislative and executive branches. And due to the steady "presidentialization" of our politics, while prime ministers in other Commonwealth countries can be ousted and replaced by the caucus, in Canada we have adopted US-style conventions to elect party leaders, leaving the caucus essentially powerless.[224]

Witness the efforts of Liberal MPs who, for months, tried to convince Justin Trudeau it was time to go. Margaret Thatcher was not given a choice by her caucus. They simply made John Major the prime minister, replacing the Iron Lady.

That is another reason why the First Past the Post system is so dangerous for Canada. A prime minister can gain 100 percent of the power with less than 50 percent public support. What if we had a right-wing populist in Canada as bad as Donald Trump? The only way to be sure such a person could not get 100 percent of the power is to fix our voting system—while there is still time. Only First Past the Post allows for such "false majority" governments, perverting the public will.

First Past the Post distorts the will of the voters, discourages true representation and perpetuates a system that blocks progress on critical issues like climate action, affordable housing and social justice. But we can change that—and Elizabeth May is at the forefront of the fight for electoral reform.

In a minority parliament, where holding the confidence of the House could come down to one or two votes, the minority party could hold the balance of power—and that means they could have the power to insist on the change we urgently need, like electoral reform and climate action.

Only when every Canadian has a voice, when the government works together instead of being intent on dividing the masses, only then will faith in our government begin to be restored. If not, I despair to think where this downward spiral away from democracy will take us. The sooner every Canadian becomes aware of the dire situation of our democracy, the sooner we can revolutionise our recovery.

Together with fourteen co-sponsors, I filed a federal petition late in 2024, demanding national drinking water regulations which include source water protection. Sadly, after receiving 2,000 signatures, government was prorogued and our petition was no longer active. We were told to start over again.

How we protest is key. How can we be most effective? We have decided to launch a national letter writing campaign to Prime Minister Mark Carney. The details of the campaign can be found at the end of this book. I hope you will support it by signing, sharing and talking it up with your network.

I am super confident in Carney's ability to lead Canada, to avoid potential for escalating conflict with the US and to ensure we don't become the 51st state. My hope is that he will apply his values to people's health and rights, embrace taking back control of our water, initiate national drinking water regulations and source water protection.

Now it is up to you, the readers. Please bring this issue to the attention of your elected officials and to your friends. Ask them to also share this important message. Share with media, celebrities or anyone who will support our message, please do so without hesitation.

To each of you, I make this offer in thanks for your advocacy. If you organize a Zoom meeting, I will happily volunteer my time to answer your questions or give a presentation. I will also donate a free copy of *Water Justice* for you to deliver to your MP and/or your MLA and ask that you maintain pressure on them to act on what they learn. Watch my website for the latest list of MPs and MLAs who have been gifted a copy of *Water Justice*. I will post current challenges and keep updating any responses I receive to the FOI requests and complaints I filed.

Please also take the time to review *Water Justice*; Amazon or Goodreads are both good platforms to share your opinions.

UNPRECEDENTED PARTNERSHIP

For the Okanagan Syilx peoples, water—sisiwɬkʷ—is sacred, connecting communities to land and all living beings. Their ancestors have carried the responsibility to protect water for future generations. Today, that responsibility is shared through an unprecedented partnership between First Nations and local governments across the Okanagan and Similkameen watersheds.

The Okanagan–Similkameen Collaborative Leadership Table unites leadership from Okanagan Syilx communities with seven municipalities and regional districts across more than 10,000 square kilometres of lakes, rivers, creeks, and aquifers. Through a historic 2024 Memorandum of Agreement, these partners have committed to co-create solutions for watershed health.

This co-leadership model, rooted in trust and mutual respect, recognizes that protecting water requires collaboration and shared governance—demonstrating that when Indigenous knowledge and local government work together, watersheds can be effectively protected for all.

Chapter 62

Whose Future Is It Now?

Young people consistently identify climate change, mental health and affordability as their top concerns. They are deeply engaged and willing to take action.

Each spring a passionate local artist in my hometown of Qualicum Beach on Vancouver Island organizes an art competition for students as part of mental health month. The theme in 2025 was titled "Have Your Say" as part of Mindfulness in May. The elementary winner, a grade seven student submitted this poster.

When asked to describe her poster, thirteen-year-old Ehnyawpaw Johdee, wrote: *Everyone needs to listen to everyone with respect.*

She is more astute than the politicians who ignore their constituents and follow their leaders. I think she summed up the exact problem. We are fed up with shouting, trying to educate politicians, trying to be heard by politicians who don't want to listen.

A grade twelve student, Julie Chambers, submitted this poem:

Have Your Say
Yes

264

No

"Maybe so"

"What do you think?"

"Agree"

"Or not so much?"

"Voice your opinion"

"Raise your hand"

"Tell me your thoughts"

"We won't judge"

"Speak your mind"

"Repeat that, we couldn't hear you."

"Maybe if you weren't so busy

Drowning in the noise of the bustling world

You could hear my words"

You say you won't judge

Then why do I hear whispers?

You tell me to speak my mind

But why do I get laughed at?

You tell me to have my say

But why do you shove it to the side?

How am I supposed to change the world like I'm told to?

When no one will acknowledge my presence?

All I want for a change

Is to have my say.

These young women represent our future, their future and their generation's feeling of hopelessness. I suggest we should lower the voting age to sixteen, as recently happened in the UK. It's not our future anymore.

I have included many stories of individuals who have become passionate about water or other environmental causes and turned into enthusiastic activists. I would like you to look closely at who they are. The majority were young, under thirty, many under twenty, when they began their missions.

When I visit elementary and high school science fairs, there is almost always at least one student attempting to purify water for drinking. I am consistently encouraged by young people advocating for a better world. We can't expect youth to be proactive unless we give them a voice and a platform for change. We must demand better. One person at a time, I believe we can affect change.

There are some people who claim that there is no point trying to remove fossil fuels from our society as India and China are going full steam ahead. Maybe they were decades ago, but not anymore. Research what is happening around the world.

CHINA - WORLD'S LARGEST RIVER HABITAT REVIVAL

China has demolished 300 dams on the Chishui River 赤水河, a major tributary of the upper Yangtze, restoring 400 kilometres of river and luring the once-vanished Yangtze sturgeon back to spawn.

Begun in 2020, the cleanup has reopened migration routes and cut small-hydro output by 90 percent, surpassing the Klamath as the world's largest river-habitat revival.

It's working. Scientists observed the first hatching of wild sturgeon in April 2025.

Chapter 63

Value of Water

The stories I share here are just a few that have crossed my path. Each of the issues I identify could easily become books of their own. I suspect my account of government inequities amounts to just a teaspoon of water in a large lake. I am sure there are thousands more that are not on my radar. I welcome readers to draw my attention to other water-related injustices.

I am not as naïve as when we founded SDWF in 1996. It can't be that politicians need to be educated; it must be that we need to convince those in power that it is simply the most cost-effective solution to protect our source waters and establish drinking water regulations. It comes down to $ and common sense. To logic. To values. Which Prime Minister Carney has written on.

I am reminded that May 10, 2005, an internal briefing note to Environment Minister Stephane Dion was published from an Access to Information request. The document states "Our failure to protect water has caught up with us." It continues: "Diseases from contaminated water cost our health system $300 million a year. Between 20 percent and 40 percent of rural wells are more contaminated than drinking water guidelines recommend." I can only imagine that these costs have compounded and accumulated much like the pollutants in our waters. Now two decades later, I estimate the costs to be at least $1B annually. It's not just the cost. It's the suffering! Not just for the people, for the environment, for our future generations. I ask Prime Minister Carney, surely, he sees the financial logic in protecting source waters and implementing national drinking water regulations. Stop the suffering as well as the monetary cost.[225]

What I did read in his book was that Mark Carney is a strong believer in markets but acknowledges they need to be regulated for there to be decent outcomes. I suggest that the same applies to safe drinking water and to AI. Maybe we can convince him to treat water as though it is an investment, which it is. An investment in decreasing demands on health care, decreasing the effects on climate change, decreasing the strain on the environment. An investment in reconciliation. An investment in future generations.

He also clarified that for a successful modern economy, there are seven essential values and beliefs: dynamism, resilience, sustainability, fairness, responsibility, solidarity and humility.

First Nations have certainly shown resilience. I like to think settler Canadians are showing solidarity. Prime Minister Carney seems to have a dynamic personality. What is left is sustainability, fairness, responsibility and humility. I hope he is prepared to contribute each of those to our democracy—they are long overdue.

I wish to be very clear here: Prime Minister Carney did not initiate the demise of our democracy, nor did he dismiss the many petitions signed by thousands of Canadians. He hasn't made empty promises about lifting BWAs or reconciliation with First Nations. But neither has he embraced and welcomed consultations with Indigenous people. He is not responsible for Canada's laws on governance; however, he needs to fix our electoral reform to avoid a Trumpite taking control of Canada. Remember, President Trump was elected by a First Past the Post electoral system.

Carney is now the only person who can make changes to reinstate our democracy, to give all Canadians a voice, to establish drinking water regulations and lead our country amid its forest fires, floods and droughts to face climate change and take logical steps that protect the well-being of future generations.

If Prime Minister Carney is serious about ending the government's irresponsible spending, if he is serious about values, he can easily cut billions. I suggest three inquiries with the ability to affect change: first into ISC spending, second into contracts awarded by First Nations with ISC funding and third to determine why funded scientific research projects have not been acted on or implemented, would be a great place to start.

I am reminded of something which happened to me years ago. I stood in a grocery store, opening a bulk food dispenser for flax seed, which was higher than me. I reached high above my head and held the plastic bag tight around the neck of the container, putting slight pressure on the lever with my other hand. Along came a dear friend I hadn't seen in ages, she tapped me gently on one shoulder and I turned my head, so pleased to see her I lessened my grip of the bag around the neck. As I leaned into her, the flax seeds began spilling all around us. I didn't realize for a few seconds what was happening. I didn't react to release the lever, which allowed the flax to keep falling. We burst out laughing as flax seeds cascaded around us.

I see the same scenario with the government looking after our tax dollars and our well-being: the money is flowing, scattering among society, but we don't have anyone addressing the need for drinking water regulations and source water protection. Our governments have not taken care of Canadians' greater good. The money is not scattered around for anyone to pick up, it is falling into bank accounts of those who have connections, the lobbyists, the corporations, engineers who receive repeated contracts to fix what they failed to build effectively the first time. And I wonder who else. No one is laughing!

While it may seem too late to protect the earth from climate change, every effort still counts. Immediate action can mitigate the worst impacts, preserving ecosystems and human life. It's crucial to reduce emissions, adopt sustainable practices and support policies that prioritize the planet's health. Above all, we must take back our water. Every step matters.

If you have questions about any aspect of effective water treatment, source water protection or any of the contaminants that threaten human health, I encourage you to research the extensive Fact Sheets available from SDWF by visiting safewater.org.

YOUTH-DRIVEN PROTESTS FORCE GOVERNMENT TO RESIGN

Bulgaria's late-2025 street protests used the same playbook as Nepal and Madagascar: Young organizers turned online networks into mass turnout, culminating in the government's resignation in December 2025.

"What this moment makes clear is that social media has not distracted Gen Z from civic life but, instead, placed political participation in the palm of their hands."

Chapter 64

The Man at the Top

December 1, 2025, nineteen months after I first submitted an FOI to CWA, and finally their response sits in my inbox. The attachment is 175 pages! I glance at the opening pages, knowing I cannot read and review this in time for this book to go to print. Seven pages in and I have confirmation of exactly what my gut told me all along. The results, including this document, will be posted on my website, but it is not cited at this time. BCAFN submitted a letter to the CWA March 31, 2021. Here I quote their words:

> Furthermore, in the key findings of the 2021 Auditor General's Report, the audit found that Indigenous Services Canada (ISC) failed to provide First Nations communities with the adequate resources and supports to ensure access to safe drinking water on reserves. While this is a troublesome but not surprising assessment, it is concerning that ECCC has indicated that despite ISC's failure to end all long- term boil water advisories by March 2021, that considerations for the development of the CWA will not encompass safe drinking water on reserves given that is a "pre-existing commitment" by the Federal Government. This contradicts ECCC commitment to working with First Nations by demonstrating that priorities have been pre-

determined for First Nations without prior consultation and engagement. These actions both undermine ECCC's commitment to reconciliation and collaborative processes with First Nations on the CWA and its "aim to help the government of Canada advance reconciliation with Indigenous peoples with respect to areas under federal jurisdiction.

It's not a pre-existing commitment in my mind. It is a flagrant pre-existing embarrassment for Canada.

Does it all come down to values? Or does it amount to deceit?

In 2021, Trudeau was Prime Minister declaring he would lift all BWAs, but in reality he established the largest funded government organization focussed on water and refused to touch the most challenging, most sensitive, most costly of all water related issues in Canada: First Nations' lack of safe drinking water.

The CWA was a commitment of the Liberal Party under Prime Minister Trudeau, during the 2019 and 2021 election campaigns. My gut was right; I believe CWA only wanted "yes" people who would not challenge the status quo.

Shameful. Disgusting. Shocking. Criminal. These are words that come to my mind. We have an Ontario Ombudsman flying into First Nations in the name of reconciliation when the man at the top appears to have absolutely no intent of reconciling. Who can blame civil servants for failing when the man at the top sets the agenda, with no intent of improving First Nations quality of drinking water. Justin Trudeau put his acting skills to good use. What an act of deceit.

So now it is up to the author of *Values*. Let's see if he can walk the talk, and put his words into action. As the federal government promises revisions of Bill C-61, once again they appear to be applying patriarchal colonizer dictation of what this Bill will contain instead of asking and listening to the Chiefs and First Nation leaders. When land rights, doctors' whistleblowing of cancers in First Nation communities, the truth and reconciliation recommendations and a host of other laws have failed to change the colonial attitude of our government, I continue to hope for much needed and significant change. Surely enough is enough!

My analysis of the records received from CWA will be posted on my website. Additionally, in late December 2025, another FOI I submitted for the same information resulted in a huge package of documents, hundreds of pages, with no comparison to what I received in response to the earlier identical FOI. I see hundreds of pages of redacted information, too many duplicate pages, correspondence citing honorarium that were not paid as promised. And one comment that led me to check with Google. AI tells me The Canada Water Agency (CWA) and Australia's Department of Climate Change, Energy, the Environment and Water (DCCEEW), along with the Australian Water Partnership, are partners in an international knowledge-sharing initiative.

This collaboration focuses on sharing expertise regarding freshwater stewardship and governance. So, our federal government are eager to partner with the only other developed country who does not have drinking water regulations. It seems to me to be they will partner with anyone rather than give Indigenous people the right to govern their own land and water. This latest information package will also be posted on my website with a summary.

INDIA HAS EXPANDED RURAL TAP WATER ACCESS

India has expanded rural tap water access from 16.7 percent of the population in 2019 to 81 percent in 2026, connecting 125 million rural households to clean, running water.

In sheer numbers, this is the biggest, fastest, and most important sanitation drive in human history.

Why has it not been more widely reported?

Chapter 65

Dear Prime Minister Carney

We need national regulations to ban all forms of pollution from entering our source waters, with significant fines for offenders.

We the citizens and owners of Canada need to take back our water, to protect it from capitalism, colonialism, and government negligence. We need to return source waters to the life- giving sustenance they provide to all life on earth.

That means we must ban corporate lobbyists, whose agendas threaten either quality or quantity of water. We must implement national drinking water regulations which are legally enforced and protect our source waters.

Every Canadian has the fundamental right to expect their federal and provincial governments protect the environment and, by extension, public health. This requires strict control of waste, toxic substances, and pollutants through robust regulations that guarantee a sustainable environment for future generations. Anything less is a betrayal of the government's most basic responsibility to its citizens.

Prime Minister Carney's climate record doesn't match his rhetoric. Canadians deserve better. They need to be planning national drinking water regulations that are based on source water protection. This fact confirmed by a UN report, January 2026, which declares the world has entered a "global water bankruptcy," impacting billions.[226] Urgent action is needed to address overuse and pollution of water resources, as many systems are irreparably damaged. With 75 percent of the global population in water-insecure regions, conflicts over water are rising. The report emphasizes the need for a fundamental reset in water management to prevent further losses and maintain social stability. Prime Minister Carney needs to reconsider his position and act to protect source waters in Canada.[227]

The value of safe drinking water, and thereby source water protection, cannot, must not, be underestimated. Harms caused by toxic chemicals are too great to ignore. Toxic chemicals in food systems impose catastrophic health costs that governments fail to address. Phthalates, bisphenols, pesticides, and PFAS cause developmental disorders, premature mortality, metabolic and circulatory diseases, creating systemic health burdens.

The annual global cost: $1.4–$2.2 trillion (2–3 percent of global GDP)—exceeding the combined profits of the world's one hundred largest companies. With phthalates alone costing $816 billion, bisphenols $609 billion, pesticides $533 billion, and PFAS $227 billion, governmental inaction on toxic chemical regulation represents an economic and public health catastrophe. Canadians and environmental organizations must pressure authorities to act, as the price of doing nothing far surpasses prevention costs.[228]

We can sit and do nothing, or we can advocate for change. We can make one attempt, and, when we get shut down, we can quit and declare the system is against us—which it is! Or we can increase our demand, time and time again, shout louder, share with more people, maintain our demands until they are met and we can say we made a difference. Only when enough people stand up and demand change will we see change.

With great appreciation to the National Council of Women for hosting the national letter writing campaign to Prime Minister Carney, and the government of Canada, with strong support from Soroptimist International and all co-sponsors of the original, prorogued federal petition.[229] Please share this link with everyone you know: https://www.ncwcanada.ca/open-letter-for-canada-to-protect-our-water

You can also find the link on the websites of some co-sponsors of the original federal petition websites.

Please sign the following letter to **Prime Minister Carney: Protect Canada's Water–Protect Our Health**. The link is also available on my website at susanblacklin.com and via many other organizations supporting this message, hosted by the National Council of Women of Canada.

An Open Letter to the Government of Canada
Protect Canada's Water — Protect Our Health

As Canadians deeply concerned about our nation's growing water crisis, we, the undersigned organizations and individuals, call on the Government of Canada to act now.

We urge immediate implementation of national, enforceable drinking water regulations that include strong source water protections — to safeguard the health, environment, and future of every person in this country.

Water Is Life — and Canada Is Falling Behind

Despite billions spent on research, too many communities continue to drink contaminated water. Scientists have found more than 50 different pesticides in Canadian tap water. Beneath our streets, ageing asbestos cement pipes are breaking down, releasing harmful fibres into drinking systems. And across the country, First Nations communities such as Neskantaga and Grassy Narrows have endured boil-water advisories for decades, despite repeated government promises and investments.

The government's decision in Budget 2025 to eliminate periodic pesticide re-evaluations would weaken existing protections even further. Shockingly, this proposal relies on a scientific study later retracted for ethical breaches: Science.org – Journal retracts Monsanto-backed study. Canadians deserve science-based policies that protect public health — not corporate influence over safety standards.

A National Failure with National Consequences

Canada is the only G7 nation without enforceable national drinking water regulations. Provinces rely on inconsistent, often weak guidelines that cannot guarantee equal protection for all residents. Even modern treatment systems cannot remove every contaminant, leaving Canadians exposed to chemicals and microbes that cause cancers, infertility, immune dysfunction, and waterborne disease.

The Global Evidence Is Clear

A new international report, Invisible Ingredients, produced by leading scientists from institutions including the Institute of Preventive Health, the Centre for Environmental Health, the University of Sussex, and Duke University, shows how synthetic chemicals — phthalates, bisphenols, pesticides, PFAS — are poisoning global ecosystems and human health. The report estimates that toxic exposures impose up to $2.2 trillion (USD) in annual health costs, and if left unchecked, could lead to 200–700 million fewer births globally by 2100.

The experts behind the study describe chemical pollution as a crisis rivaling climate change — yet Canada continues to lag behind the World Health Organization, the European Union, and the United States in setting enforceable standards.

The Cost of Inaction

The numbers are staggering, but the lesson is simple: ***Prevention is far cheaper than cleanup.***

When we account for the combined costs of waterborne illnesses, pesticides, and industrial pollution, the economic and health costs of inaction become indefensible. Prevention — through strong source water protections and consistent enforcement — will save lives, reduce healthcare spending, and protect future generations.

Our Call to Action

We call on the Government of Canada to:

- Adopt legally enforceable national drinking water regulations applicable in every province and territory.
- Include comprehensive source water protection as a central pillar of these regulations.
- Ensure transparency, accountability, and public access to water quality data in all jurisdictions.

Clean water is a fundamental human right. It transcends politics, geography, and ideology. Protecting it is the responsibility of all levels of government — and the duty of every Canadian to demand. We stand together—environmental advocates, scientists, healthcare professionals, community leaders, and citizens—united in calling for leadership that matches the scale of this crisis.

Safe drinking water cannot wait. The time for national action is now.

Signed,

～

Thank you for reading my book, I welcome your thoughts and suggestions on how together we can make our world a better place. Contact me at blacklinsusan@gmail.com to schedule a Zoom presentation to your group.

Acknowledgments

Self-publishing is a huge learning curve, a daunting task and consumed my life this past year. When I was mentally exhausted, from so many challenges along the way, two angels appeared in my life, working in unison together they corrected and polished this book, I greatly appreciated their efforts. Christina Myers stepped up to proofread, catching many idiosyncrasies and some bigger ones. Jennifer Sommersby then designed the cover and formatted the entire book. Together they gave *Water Justice* the professional finish it deserved. I am hugely indebted and thankful to them both for their highly professional and talented approach.

My sincere gratitude goes to Nicole Hancock, the passionate executive director of SDWF, whose keen eye for detail, encyclopedic knowledge of all things water, meticulous fact-checking and thorough editing made this work possible.

To all the readers of *Water Confidential* for bringing countless critical issues to my attention, for all your tips, news links and "have you heard about this," thank you, this is your book.

I give big thanks to Penny Rankin, President of the National Council of Women of Canada; Carell Wingrave, Vice President of Soroptimist International Canada Online club and Advocacy Chair for Western Canada Region; and Suzanne Heron, Membership Chair for Soroptimist International Canada Online Club, for their continued support and encouragement.

To Dr. John O'Connor, the busiest person I know, who repeatedly found time to share his thoughts, suggestions and appreciation, which motivated me to write from the beginning.

One could easily write another book replacing the word "water" with "forests." The real crisis in Canadian forestry isn't tariffs, its over-cutting, habitat destruction, declining revenues, and escalating flood and fire risks. It's time for a new Forest Act. The same urgent plea could be made substituting water with housing costs, food insecurity, homelessness, the opioid epidemic, mental health issues, our failing healthcare system, failing education system, immigration challenges or systemic racism, above all our failing democracy and the ever-increasing disparity between the elite and the poor.

Anyone working in a charity addressing all critical issues, (who does not receive government financing) is becoming so fraught, no day is long enough, the issues keep coming and don't keep up with the donations. Thank you to each one of you for striving to make the world a better place.

I also thank all the journalists who fight tirelessly to deliver sound, unbiased news that informs and educates Canadians, often against overwhelming odds, and with similar struggles to attract funding as charities.

I salute the scientists who conduct their research knowing full well it may never be applied for the greater good, yet they persist anyway.

I thank the readers of *Water Justice* in advance, hoping fervently that they will become fierce advocates for drinking water regulations by sharing this book with elected officials, friends and family, sparking difficult, but necessary, conversations, and by signing the open letter to the Government of Canada, and encouraging others to join them.

I thank my beta readers who identified that I had to focus on the facts and curb my anger. I hope I have shown my readers why I am angry. I have great respect for author, feminist and activist Soraya Chemaly, whose best-known book, *Rage Becomes Her: The Power of Women's Anger*, has motivated and inspired me. "Women and girls are more likely to feel anger about injustice, powerlessness and other people's irresponsibility." Women and girls are also most likely to be affected by unsafe drinking water. Thank you to all those who stood by me with unending positive encouragement.

Finally, but never least, I give profound thanks to my partner, Roelof, for his steadfast support and encouragement through every step of writing this book: for every meal prepared, every task undertaken, every sacrifice made to give me the time and space to write this past year. I could not have done this without you.

Appendix One: Letter to All Saskatchewan First Nation Chiefs from Capital Manager, Indigenous Services Canada (ISC) Saskatchewan Region

Indigenous Services
Canada

Services aux
Autochtones Canada

1827 Albert Street
Regina, SK   S4P 2S9

Our file - Notre reference
GDOCS 91980710

March 16, 2021

All Saskatchewan First Nation Chiefs

Dear Chief,

The Government of Canada recognizes the importance of operations and maintenance (O&M) in ensuring First Nations communities have sustainable infrastructure. That is why we have committed to new investments that will significantly increase the annual funding provided to support the operations and maintenance of water and wastewater systems in First Nations communities.

In March 2021, First Nations will receive a one-time supplement to water and wastewater O&M funding for the current fiscal year, ending on March 31, 2021. These increases mean that 100% of water and wastewater operations and maintenance costs—up from 80%—will be covered based on the operations and maintenance funding formula.

First Nations will have the same flexibility as they have now in determining how this money is spent towards the operations and maintenance of their water and wastewater systems. There are no new reporting requirements, and no new expectations regarding the costs First Nations are responsible for in relation to operations and maintenance of water and wastewater systems.

Funding transfers will be completed by the end of March 2021. First Nations presently funded by Indigenous Services Canada under Grant or Block arrangements can expect to receive a funding amendment shortly. Funding transfers to First Nations under regular funding arrangements will be processed through a Notice of Budget Adjustment (NOBA).

This in-year allocation does not reflect future O&M allocations for water and wastewater systems. For future years, allocations will continue to cover 100% of formula-based O&M costs for water and wastewater systems, however a new allocation methodology is being developed. This new allocation methodology will include updating the O&M formula to better account for O&M costs in fiscal years 2021-2022 and beyond. It will also include stabilization of funding for capacity building initiatives such as the Circuit Rider Training Program.

These new investments provide an opportunity to make substantive progress towards Asset Management Reform – providing a predictable funding stream that

allows for strategic decision-making built upon detailed asset information. Indigenous Services Canada continues to work with First Nation partners to reform our Operation and Maintenance policy into an Asset Management Policy.

If you have any questions, please contact myself. I can be reached at derek.tuchscherer@canada.ca or (306) 540-2265.

Sincerely,

Digitally signed by tuchscherer, derek
DN: C=CA, O=GC, OU=ISC-SAC,
CN="tuchscherer, derek"
Reason: your signing reason here
Location: your signing location here
Date: 2021.03.16 11:22:35-06'00'
Foxit PhantomPDF Version: 10.1.0

Derek Tuchscherer,
Capital Manager
ISC Saskatchewan Region

cc: Kevin Lysak, Associate Director North Budget Centre
        Rhonda Dawiskiba, Associate Director Central Budget Centre
        Rita Kinequon, Associate Director South Budget Centre

Appendix Two: Water First's Outcomes Funded by Indigenous Services Canada (ISC) 346

Outcomes:

ISC funding will directly support approximately 50% of Water First's administrative wage costs over the next 5 years, which currently includes 10 management employees at various levels. Our high-level administrative objectives in the next 5 years are:Yearly alumni event starting in 2024/25, bringing interns together for training and networking.

- Maintaining and enhancing staffing supports, while aligning with non-profit best practices;
  - Retaining and expanding when appropriate, an experienced and professional human resources team, currently at 1.1 full-time equivalents (FTEs);
  - Supporting, retaining and expanding when appropriate, Water First's overall staffing base, currently at 40 employees, and based on trends and current plans the staffing base is anticipated to reach up to 60 employees within 5 years;
- Optimizing Water First's accounting systems for continued growth;
  - Retaining and expanding when appropriate, an experienced and professional financial management team, currently at 2.5 full-time equivalents (FTEs).
- Organizational management models and tools optimized and adapted for ongoing expansion of Water First;
  - Hiring and retaining a Sr. Director of Internal Operations in 2024/25;
  - Hiring and retaining a Sr. Director of External Operations in 2025/26;
  - Optimizing success metrics, data collection and analysis, both internally and externally, adjusting strategy where appropriate, and continuing to deliver industry-leading programming.

Suggested Discussion Topics

Applicable to individual readers, book clubs, high school social studies or science curriculum support and for university students studying Indigenous studies and/or social sciences. I recommend discussions focus on the issues and not on the political parties to avoid divisive conversations.

What shocked you most in *Water Justice*?

International

- That Canada is the only country in G7 who does not have national Drinking Water Regulations
- That EU and UK drinking water regulations have more stringent standards for multiple pesticides
- Privatization of Thames Water in London and repercussions for UK citizens
- Plight of people in places like Brazil or Hawaii at hands of large corporations

Indigenous Issues

- That First Nations are under federal jurisdiction whereas urban and rural settler communities fall under provincial legislation?
- That ISC do not have documentation to prove how funding is allocated by First Nations
- Bill C-61
- The multiple assaults on Indigenous communities and their inability to fight them all
- The continuing saga of Swan Hills Toxic Waste site and affected communities, after decades of challenges, still not having safe drinking water

Health Concerns

- How waterborne illness can wait decades to show
- Why the Walkerton Inquiry has not been applied to First Nations in Ontario or all of Canada
- The need to educate physicians
- That health costs do not appear to be of value or concern as they are not included as liabilities against the profits of resource rich corporations
- Why the Dryden Paper Mill has not been banned from releasing effluent laden with mercury

Federal/Provincial Jurisdiction

- That source waters in Canada fall under provincial laws and are not protected
- Bills - Bill C-61, Bill 5, Bill 54, Bill C 5, Bill 15
- The differences in provincial drinking water guidelines
- The various DWAs
- The number of oil spills or corporate infractions and the small fines

Is our democracy failing?

- Federal
- Municipal
- Provincial

Who is really running Canada?

- The power of corporations and their lobbyists
- The varied fines by provinces
- The low fines charged to corporations owned by multi billionaires
- Government contracts awarded without competition
- That research funding appears to favour documenting what is polluting our source waters or profiteering from developed business products
- That little, if any, research is prioritized and applied to improve water treatment systems
- Difficulty obtaining responses to FOI requests
- Governments are disinterested in those initiating or signing petitions and they are not acting on petitions

Education

- SDWF offers sponsored school kits for students in grades four through twelve.
- Will you encourage schools near you to teach the SDWF Operation Water Drop school programs?
- The national water education program being reinvented by AquaAction from the CAAF.
- AI being rammed into our society with no plan for the true costs to society

Charities

- The lack of government oversight for charities
- Laws governing charities

- Government agencies or organizations registered as charities
- The lack of advocacy of some charities

Threats to our drinking water:

- Asbestos cement pipes
- Agriculture
- AI
- Algae
- Big Oil
- Bottled Water
- Chemicals
- Climate change
- Coal Mining
- First Nations not allowed to test their own waters
- Fracking
- General Motors
- Gold Mining
- Nuclear
- PFAS
- Pipelines
- Sewage
- Scientists controlled for what they test and what they disclose
- Sewage sludge as fertilizer
- Swan Hills Toxic Waste Site
- Transport Canada

Advocacy

- *Water Justice* was not an easy read, are you glad you read it?
- Will you sign the petition to Take Back Our Water
- What are you able to do as an individual?

- What groups do you belong to who would benefit from knowing this information?
- Who will you recommend to read Water Justice?
- Will you write a review on Amazon?
- Will you share the petition, to Take Back Our Water

Susan Blacklin is available to answer readers' questions or give presentations via Zoom. She asks, if possible, that you give tax-deductible donations in appreciation to either SDWF (which she helped found) or to KOW. Susan will send any group that is organizing a Zoom and making a donation a free copy of *Water Justice* to give to their MP or MLA.

Will you arrange a Zoom presentation with the author?

Will you give a copy of *Water Justice* to your MP and/or your MLA?

Will you give Susan the names of your MP and/or MLA to whom you pass on a free copy of *Water Justice* to be noted on her website?

Recommended Learning

If you wish to understand any scientific issues around water, visit: https://www.safewater.org/fact-sheets

If water is a top priority for you, I highly recommend you subscribe and follow news from:

Watershed Sentinel: https://watershedsentinel.ca
The Narwhal: https://thenarwhal.ca/
Vancouver Island Water Watchers: https://www.vancouverislandwa terwatchcoalition.ca
Great Lakes Ecoregion: https://www.greatlakesecoregion.org/

If you find it hard to believe that Canada is losing its democracy, I suggest you read Andrew Coyne's *Crisis of Canadian Democracy.*

If you like reading accurate, well-researched, current news, from independent journalists, please subscribe to:

Reuters: https://www.reuters.com/
Dogwood: https://www.dogwoodbc.ca/

The Maple: https://www.readthemaple.com
Ricochet Media: https://ricochet.media
The Tyee: https://thetyee.ca
The Walrus: https://thewalrus.ca
Canadian Affairs: https://www.canadianaffairs.news/about/

If you like to learn what is going on in the world that CBC fails to report, follow either or all of the following:

Al Jazeera: https://www.aljazeera.com
BBC News: https://www.bbc.com/news

If you haven't met Canada's equivalent of Bernie Sanders, I encourage you to meet and follow:

Charlie Angus: https://www.charlieangus.ca

If you dislike all the negative news in the world, learn about all the positive things happening, subscribe to:

Fix the News: https://fixthenews.com/

If you like to follow the footsteps of Indigenous people's lives, join **APTN** at https://www.aptnnews.ca.

Documentaries to watch on YouTube

Canada's Dark Secret: https://www.youtube.com/watch?v=peLd_jtMdrc

Shannen and her dream, by Alanis Obomsawin, Hi-Ho Mistahey!

Paul Whitehouse – Our Troubled Rivers
Thames Water – Inside the Crisis

Blue Gold: Water as a Human Right - https://www.youtube.com/watch?v=B1gPCwErTr4

To learn how people in Bolivia rose to protect their water, how the world is running out of fresh water and there is no environmental crisis as great as the commodification of the world's water supply by giant corporations: Maude Barlow, national chairperson of The Council of Canadians and author of Blue Gold, describes the movement to guarantee a water-secure future based on conservation, equity and the public good.

We All Live Downstream: A Clean Water Action Podcast, https://www.buzzsprout.com/1936739

For the true history of colonialism that you were not taught in school, watch this documentary**:** https://www.anglican.ca/primate/tfc/drj/doctrineofdiscovery

The Theory of Water will teach you how colonialism and capitalism have abused Indigenous culture. I highly recommend reading this sensitive, gentle, eye-opening book by Leanne Betasamosaki.

If you prefer dystopian novels, I highly recommend three:

Watershed by author Doreen Vanderstoop. Keep in mind, while Doreen wrote it as dystopia and two years prior to my publishing *Water Confidential*, it is an uncannily accurate prediction of what is to come unless we take back control of our water.

Grey Wolf and *Black Wolf* by Louise Penny, two novels, a two-part political thriller about water, environmental disaster, and conspiracy. This fiction was uncannily real for me.

Endnotes

1 Aaron Atcheson, Hugh Hunt, Owen Atcheson, "Legal Column: Water Blockage - Barriers to greater use of grey water systems" Water Canada, September/October 2025, https://emagazine.watercanada.net/?pid=ODk8907897&p=11&v=1.6.

2 National Collaborating Centre for Infectious Diseases, "*Helicobacter pylori*," June 15, 2023, https://nccid.ca/debrief/helicobacter-pylori.

3 "Helicobacter pylori"

4 Philippe Willems, Janie de Repentigny, Galeb M. Hassan, Sacha Sidani, Genevieve Soucy, Mickael Bouin, "The Prevalence of *Helicobacter pylori* Infection in a Quaternary Hospital in Canada," *Journal of Clinical Medicine Research*, vol. 12, no. 11, 2020, https://pmc.ncbi.nlm.nih.gov/articles/PMC7665871.

5 Health Canada, "Guidance on waterborne pathogens in drinking water," September 2, 2022, https://www.canada.ca/en/health-canada/services/environmental-workplace-health/reports-publications/water-quality/guidance-waterborne-pathogens-drinking-water.html.

6 First Nations Health Authority, "Monthly Drinking Water Advisories in First Nations Communities in BC," January 2026,

www.fnha.ca/Documents/Drinking-Water-Advisory-Monthly-Summary.pdf

7 Government of Canada, "Ending long-term drinking water advisories," October 17, 2025, https://www.sac-isc.gc.ca/eng/1506514143353/1533317130660.

8 Peter Zimonjic, "First Nations leaders want drinking water bill within 100 days of Parliament's return." CBC News, May 12, 2025, https://www.cbc.ca/news/politics/indigenous-leaders-priorities-100-days-1.7532748.

9 Andrea Hill, Ryan Kessler, Priscilla Wolfe, Michael Wrobel, "First Nations workers in Saskatchewan sacrifice wages, vacation to run underfunded water systems," APTN National News, February 25, 2021, https://www.aptnnews.ca/national-news/wages-first-nations-water-systems-saskatchewan.

10 Blacklin, *Water Confidential: Witnessing Justice Denied - The Fight for Safe Drinking Water in Indigenous and Rural Communities in Canada.*

11 Stephan Gabos, Michael G. Ikonomou, Donald Schopflocher, Brian R. Fowler, Jay White, Ellie Prepas, Dennis Prince, Weiping Chen, "Characteristics of PAHs, PCDD/Fs and PCBs in sediment following forest fires in northern Alberta," *Chemosphere*, vol. 43, no. 4–7, May 2001, https://doi.org/10.1016/s0045-6535(00)00424-0.

12 Government of Canada, "Wildfire Smoke Fine Particulate Matter PM2.5 - 72h Hourly Maps at Ground Level - 12 UTC - Prairies," November 15, 2025, https://weather.gc.ca/firework/firework_anim_e.html?type=pa&utc=12&region=prairies.

13 Chastin Martel, "Keepers of the Water against moving water between basins in Northern Alberta," The Lakeside Leader, July 22, 2025, https://www.lakesideleader.com/keepers-of-the-water-against-moving-water-between-basins-in-northern-alberta.

14 Keepers of the Water, "Programs," November 14, 2025, https://www.keepersofthewater.ca/programs.

15 Natasha Bulowski, "Federal ministers summoned over Fort Chipewyan contamination scandal," CTV News, October 23, 2024,

https://www.ctvnews.ca/edmonton/article/federal-ministers-summoned-over-fort-chipewyan-contamination-scandal.

16 Bulowski, "Federal ministers summoned."

17 Bulowski, "Federal ministers summoned."

18 Bulowski, "Federal ministers summoned."

19 The Canadian Press, "'Race did not play a part': Watchdog clears officers in arrest of Alberta chief," APTN News, December 20, 2024, https://www.aptnnews.ca/national-news/race-did-not-play-a-part-watchdog-clears-officers-in-arrest-of-alberta-chief.

20 Jack Farrell, "Alberta's plan to manage inactive oil wells now leaves taxpayers off the hook," CBC News, April 4, 2025, https://www.cbc.ca/news/canada/edmonton/alta-oil-wells-1.7502028.

21 Kyle Bakx, "'A lost opportunity': Alberta gives back $137M to Ottawa in unspent funds to clean up inactive wells," CBC News, September 17, 2024, https://www.cbc.ca/news/canada/calgary/alberta-orphan-wells-inactive-decommision-1.7324701.

22 Ben Parfitt, "A Sprawling BC Community Is Set to Lose Millions Owed by Oil Firm," The Tyee, November 18, 2025, https://thetyee.ca/News/2025/11/18/BC-Community-Set-Lose-Millions-Owed-Oil-Firm.

23 United States Environmental Protection Agency, "Climate Adaptation and Source Water Impacts," January 10, 2025, https://www.epa.gov/arc-x/climate-adaptation-and-source-water-impacts.

24 Watershed Sentinel staff, "Alberta's Emergency: Auditor blows the whistle on water mismanagement," Watershed Sentinel, October/November 2024, https://watershedsentinel.ca/wp-content/uploads/2025/01/wsoctnov2024-web.pdf.

25 Joel Dryden, "Alberta to launch 'unprecedented' water-sharing negotiations Thursday amid drought fears," CBC News, January 31, 2024, https://www.cbc.ca/news/canada/calgary/alberta-water-sharing-negotiations-rebecca-schulz-old-man-river-1.7100450.

26 RAVEN, "Athabasca Chipewyan First Nation: Restoring the River, Restoring Justice," September 22, 2025, https://raventrust.com/campaigns/acfn.

27 Joel Dryden, "Landowners, mayors divided over coal project

exploration approval in Rockies," CBC News, May 17, 2025, https://www.cbc.ca/news/canada/calgary/coal-alberta-aer-blair-painter-ron-davis-crowsnest-1.7537704?cmp=rss.

28 Dryden, "Landowners, mayors divided."

29 Kai Nagata, "How do we democratize a corporate resource colony?" Dogwood BC, April 23, 2021, https://www.dogwoodbc.-ca/news/how-do-we-democratize-a-corporate-resource-colony.

30 Michael Kosnett, "A Review of Human Health Impacts of Selenium in Aquatic Systems: A report submitted to the International Joint Commission by the Health Professionals Advisory Board," International Joint Commission, July 24, 2020, https://ijc.org/sites/default/files/2020-09/HPAB_SeleniumHealthReview_2020.pdf.

31 Health Canada, "Selenium and its compounds - information sheet," September 14, 2023, https://www.canada.ca/en/health-canada/services/chemical-substances/fact-sheets/chemicals-glance/selenium-compounds.html.

32 Yangzhuo He, Yujia Xiang, Yaoyu Zhou, Yuan Yang, Jiachao Zhang, Hongli Huang, Cui Shang, Lin Luo, Jun Gao, Lin Tang, "Selenium contamination, consequences and remediation techniques in water and soils: A review," *Environmental Research*, vol. 164, no. 4–7, 2018, https://www.sciencedirect.com/science/article/pii/S0013935118301075.

33 Health Canada, "Selenium and its compounds."

34 Vinceti M, Crespi CM, Bonvicini F, Malagoli C, Ferrante M, Marmiroli S, Stranges S. The need for a reassessment of the safe upper limit of selenium in drinking water. Sci Total Environ. 2013 Jan 15;443:633-42. doi: 10.1016/j.scitotenv.2012.11.025. Epub 2012 Dec 7. PMID: 23220755. https://pubmed.ncbi.nlm.nih.gov/23220755/

35 Alexa St. John, "Renewables become biggest source of electricity globally for 1st time, beating coal," CBC News, October 7, 2025, https://www.cbc.ca/news/science/solar-wind-renewables-coal-electricity-1.7653234?mkt_tok=Nzc0LVNITy0yMjg
AAAGdbTGOajHXTAg-jGWSHfrZdR2FZAJ8DDl5YDXSuFL21u5lt
IQD17QctzjXTQJ6m6JV_5F_2LzLrbxEALFbunHYQLG8Bry2zr
Xhreme9erSyaM

36 Jacqueline Ronson, "Bundle up for politics season," *The Narwhal,* October 10, 2025, https://thenarwhal.ca/newsletter-fall-politics-season.

37 Hosea, Leana and Ferguson, Juliet. "UK and Europe's hidden landfills at risk of leaking toxic waste into water supplies." *The Guardian.* Dec. 2, 2025. https://www.theguardian.com/environment/2025/dec/02/uk-europe-hidden-landfill-leaking-toxic-waste-water-supplies.

38 Ainslie Cruickshank, "A portrait of pollution around Canada's busiest port," *The Narwhal*, May 21, 2024, https://thenarwhal.ca/bc-burrard-inlet-pollution.

39 Dryden, "Landowners, mayors divided."

40 Hannah Moffatt, Sylvia Struck, "[ARCHIVED] Water-borne Disease Outbreaks in Canadian Small Drinking Water Systems," National Collaborating Centre for Environmental Health, November 15, 2011, https://ncceh.ca/resources/evidence-reviews/archived-water-borne-disease-outbreaks-canadian-small-drinking-water.

41 Kelsey-Sugg, Anna, and Zajac, Bec. "Australian PFAS guidelines for drinking water criticised as US, EU take significant steps on 'forever chemicals'" ABC News Australia, August 18, 2024. https://www.abc.net.au/news/2024-08-13/pfas-drinking-water-guidelines-australian-us-standards-differ/104119668.

42 Alberta Energy Regulator, "Emerging Resources - Lithium," June 2025, https://www.aer.ca/data-and-performance-reports/statistical-reports/alberta-energy-outlook-st98/emerging-resources/emerging-resources-lithium.

43 Robert Shewchuk, "LithiumBank Resources Receives License for Well at Boardwalk Lithium Project, Alberta, Canada," Junior Mining Network, May 16, 2024, https://www.juniorminingnetwork.com/junior-miner-news/press-releases/3128-tsx-venture/lbnk/161004-lithiumbank-receives-license-for-well-at-boardwalk-lithium-project-alberta-canada.html.

44 Water Canada, *Top 50 Canadian Water Projects: Inaugural Issue 2024*, https://emagazine.watercanada.net/?pid=ODg8806481&v=4.9.

45 Sarah Law, "Pikangikum First Nation takes federal government to court over lack of water, wastewater infrastructure," CBC News, May 14, 2025, https://www.cbc.ca/news/canada/thunder-bay/pikangikum-first-nation-water-wastewater-1.7533789.

46 Water Canada, "Muskoday First Nation celebrates opening of new water treatment plant," August 15, 2024, https://www.watercanada.net/muskoday-first-nation-celebrates-opening-of-new-water-treatment-plant.

47 Canada, Indigenous Services Canada, 2025, https://www.susanblacklin.com/s/Muskoday-Release-Package-A-2025-00032.pdf.

48 CBC News, "Muskoday First Nation declares emergency over water supply," July 27, 2016, https://www.cbc.ca/news/canada/saskatchewan/muskoday-first-nation-emergency-1.3698261.

49 Water Canada, "Muskoday First Nation celebrates."

50 Sarah Law, "Grassy Narrows First Nation sees start of mercury treatment facility - decades after river was poisoned," CBC News, March 6, 2025, https://www.cbc.ca/news/canada/thunder-bay/grassy-narrows-mercury-care-home-construction-1.7476038.

51 Water Canada, "Grassy Narrows First Nation Ends All Long-Term Drinking Water Advisories," October 7, 2020, https://www.watercanada.net/grassy-narrows-first-nation-ends-all-long-term-drinking-water-advisories.

52 Water Canada, "Grassy Narrows First Nation Ends."

53 "Grassy Narrows says it's 'under threat' from gold mining, nuclear waste," The Winnipeg Sun, July 12, 2025, https://winnipegsun.com/uncategorized/grassy-narrows-says-its-under-threat-from-gold-mining-nuclear-waste.

54 Indigenous Watchdog, "'The threat of more poison in our water': A gold mining firm plans to discharge wastewater upstream from Grassy Narrows," July 10, 2025, https://www.indigenouswatchdog.org/update/the-threat-of-more-poison-in-our-water-a-gold-mining-firm-plans-to-discharge-wastewater-upstream-from-grassy-narrows.

55 Sarah Law, "Mercury poisoning near Grassy Narrows First Nation worsened by industrial pollution, study suggests," CBC News,

May 23, 2024, https://www.cbc.ca/news/canada/thunder-bay/grassy-narrows-first-nation-methylmercury-study-1.7211750.

56 Eirwen Williams, "'War Is Brewing in Alaska Because of the US': Native Tribes Sue Army Over Giant Gold Mine Threatening Sacred Lands and Survival," Sustainability Times, April 13, 2025, https://www.sustainability-times.com/climate/war-is-brewing-in-alaska-because-of-the-us-native-tribes-sue-army-over-giant-gold-mine-threatening-sacred-lands-and-survival.

57 Indigenous Services Canada, "Neskantaga First Nation: Water treatment system upgrade and expansion complete, deficiencies being addressed," August 23, 2024, https://www.sac-isc.gc.ca/eng/1614887856664/1614887885919.

58 Sarah Law, "30 years under longest boil-water advisory in Canada, Neskantaga First Nation pushes for new treatment plant," CBC News, February 11, 2025, https://www.cbc.ca/news/canada/thunder-bay/neskantaga-30-years-boil-water-advisory-1.7454756.

59 Law, "30 years under longest boil-water."

60 Leanne Sanders, "As governments push development agenda, First Nations are dealing with a 'hidden genocide', says Neskantaga chief," APTN News, July 9, 2025, https://www.aptnnews.ca/featured/as-governments-push-development-agenda-first-nations-are-dealing-with-a-hidden-genocide-says-neskantaga-chief.

61 Ombudsman Ontario, "Help for Indigenous people," November 11, 2025, https://www.ombudsman.on.ca/en/help/indigenous-people.

62 Government of Canada, "Infographic: Inequalities in infant mortality in Canada," April 24, 2019, https://www.canada.ca/en/public-health/services/publications/science-research-data/inequalities-infant-mortality-infographic.html.

63 Yellowhead Institute, "Calls to Action Accountability: A 2023 Status Update on Reconciliation," December 2023, https://yellowheadinstitute.org/wp-content/uploads/2023/12/YI-TRC-C2A-2023-Special-Report-compressed.pdf.

64 Michael Tjepkema, Tracey Bushnik, Evelyne Bougie, "Life expectancy of First Nations, Métis and Inuit household populations in

Canada," December 18, 2019, https://www150.statcan.gc.ca/n1/pub/82-003-x/2019012/article/00001-eng.htm.

65 Oxfam Canada, "Why Extreme Inequality Must Be at the Top of the Agenda at Davos 2025," January 20, 2025, https://www.oxfam.ca/story/why-extreme-inequality-must-be-at-the-top-of-the-agenda-at-davos-2025.

66 Natural Sciences and Engineering Research Council of Canada, "NSERC Investments: Water-Related Research," NSERC, July 1, 2025, https://www.nserc-crsng.gc.ca/_doc/NSERC-CRSNG/FactSheets/Water_Factsheet_EN.pdf.

67 Alain Marcoux, Geneviève Pelletier, Christelle Legay, Christian Bouchard, Manuel Rodriguez, "Behavior of non-regulated disinfection by-products in water following multiple chlorination points during treatment," *Science of The Total Environment*, vol. 586, 2017, https://www.sciencedirect.com/science/article/abs/pii/S0048969717303078.

68 Marcoux, Pelletier, Legay, Bouchard, Rodriguez, "Behavior of non-regulated disinfection by-products."

69 Global Institute for Water Security, "Working for Our Water Future," November 15, 2025, https://water.usask.ca/research/index.php.

70 Treasury Board of Canada Secretariat, "Confidences of the Queen's Privy Council for Canada (Cabinet confidences)," April 30, 2014, https://www.canada.ca/en/treasury-board-secretariat/services/access-information-privacy/privacy/confidences-queen-privy-council-canada-cabinet-confidences.html.

71 *Babcock v. Canada (Attorney General)*, (2002), 2002 SCC 57 at para. 18, https://decisions.scc-csc.ca/scc-csc/scc-csc/en/item/1998/index.do.

72 Canada Water Agency, "The Government of Canada strengthens freshwater protection through 65 community projects," February 14, 2025, https://www.canada.ca/en/canada-water-agency/news/2025/02/government-canada-strenghtens-freshwater-protection-through-65-community-projects.html.

73 Canada, Environment and Climate Change Canada, Canada

Water Agency, 2024, https://www.susanblacklin.com/s/A-2024-00184-CWA-Release-Oct-21-2025.pdf.

74 Amnesty International, "Extraction Extinction: Why the lifecycle of fossil fuels threatens life, nature, and human rights," November 12, 2025, https://www.amnesty.org/en/documents/pol30/0438/2025/en.

75 Climate Action Network International, "Fossil of the Day at COP30," November 30, 2025, https://climatenetwork.org/fossil-of-the-day-at-cop30.

76 Brandi Morin, "Alberta First Nations denounce plans to 'treat and release' oil sands tailings," Ricochet Media, October 27, 2025, https://ricochet.media/indigenous/alberta-first-nations-denounce-plans-to-treat-and-release-oil-sands-tailings.

77 Pacific Institute, "Announcement: Pacific Institute Analysis Finds Surge in Reported Water-Related Violence," November 11, 2025, https://pacinst.org/announcement/pacific-institute-analysis-finds-surge-in-reported-water-related-violence.

78 New/Mode, "Stop Bill 5," November 29, 2025, https://win.newmode.net/waterwatchers/stopbill5.

79 Alberta Municipalities, "Preliminary Analysis of Bill 54: Election Statutes Amendment Act, 2025," May 1, 2025, https://www.abmunis.ca/system/files?file=2025-05/ABmunis%20Preliminary%20Analysis%20of%20Bill%2054%2C%20Elections%20Statutes%20Amendment%20Act%2C%202025.pdf.

80 Rob Shaw, "Rob Shaw: NDP's infrastructure fix risks derailing Indigenous reconciliation," Business in Vancouver, May 7, 2025, https://www.biv.com/news/commentary/rob-shaw-ndps-infrastructure-fix-risks-derailing-indigenous-reconciliation-10625264.

81 BC Green Party, "Thank You for Defending Democracy, Indigenous Rights, and Environmental Protections in B.C.," BC Greens, May 17, 2025, https://bcgreens.ca/petitions/stop-bill-15/thank-you.

82 Dogwood, "Bill 15: A blank cheque for unchecked power," Dogwood BC, May 22, 2025, https://www.dogwoodbc.ca/news/bill-15-blank-cheque.

83 Kai Nagata, "PRGT: The President's Republican Gas Tentacle," Dogwood BC, July 23, 2025, https://www.dogwoodbc.ca/news/prgt-presidents-republican-gas-tentacle.

84 Xavi Richer Vis, "Who gets to talk to Carney? Natural resource lobbyists, not environmentalists," The Narwhal, October 16, 2025, https://thenarwhal.ca/carney-natural-resource-lobbying.

85 "Canada has legal duty to provide safe water, housing to remote First Nations, federal judge rules." December 8, 2025. Vancouver Island Water Watch Coalition, https://www.vancouverislandwater watchcoalition.ca/canada-has-legal-duty-to-provide-safe-water-housing-to-remote-first-nations-federal-judge-rules/.

86 Tax-Filer Empowerment Canada, "Members," November 15, 2025, https://tfec-acdr.ca/members.

87 Tax-Filer Empowerment Canada, "Doing your taxes shouldn't be taxing." July 13, 2022, https://tfec-acdr.ca.

88 Office of the Commissioner of Lobbying of Canada, "Registration - Consultant Tax-Filer Empowerment Canada / Andrew Treusch, Consultant," August 29, 2025, https://lobbycanada.gc.ca/app/secure/ocl/lrs/do/vwRg?cno=377658&regId=949149.

89 C.D. Howe Institute, "Andrew Treusch, Commissioner of Revenue And Chief Executive Officer, Canada Revenue Agency," January 29, 2025, https://cdhowe.org/events/andrew-treusch-commissioner-revenue-and-chief-executive-officer.

90 John Cairns, "Sask Party ripped for contract with lobby firm linked to Trump," SaskToday.ca, May 14, 2025, https://www.sasktoday.ca/provincial-news/sask-party-ripped-for-contract-with-lobby-firm-linked-to-trump-10661726.

91 Darren Major, "Industry minister urges Competition Bureau to conduct followup study of grocery prices," CBC News, January 30, 2024, https://www.cbc.ca/news/politics/industry-minister-competition-bureau-grocery-prices-inflation-1.7098189.

92 Water Canada, "Major federal investments into water literacy campaign for Canada," July 25, 2025, https://www.watercanada.net/major-federal-investments-into-water-literacy-campaign-for-canada.

93 AquaAction, "Empowering the next generation of water leaders," November 15, 2025, https://aquaaction.org/en/our-programs.

94 AquaAction, "Tap into the AquaNation," November 29, 2025, https://aquaaction.org/en/portfolio/tech-solutions.

95 Water Canada, "Major federal investments."

96 Environment and Climate Change Canada, "The Government of Canada invests over $3.3 million in nine grassroots, community projects to curb greenhouse gas emissions and fight climate change," June 7, 2024, https://www.canada.ca/en/environment-climate-change/news/2024/06/the-government-of-canada-invests-over-33-million-in-nine-grassroots-community-projects-to-curb-greenhouse-gas-emissions-and-fight-climate-change.html.

97 Mahad Arale, "Canada launches emissions reduction fund from Volkswagen fine," Reuters, September 17, 2020, https://www.reuters.com/sustainability/canada-launches-emissions-reduction-fund-volkswagen-fine-2020-09-17.

98 Arale, "Canada launches emissions reduction."

99 National Infrastructure Assessment Report 1, Canadian Infrastructure Council. "Building Foundations for Tomorrow." https://canadianinfrastructurecouncil.ca/national-infrastructure-assessment#waterAndWastewater

100 Righting Relations, "Adult Education for Social Change," November 30, 2025, https://rightingrelations.org.

101 Fair Funding for Nonprofits, "Fair Funding for Nonprofits: Transforming federal funding to amplify community impact," November 30, 2025, https://fair-funding-nonprofits.ca/?utm_content=buffer15977&utm_medium=social&utm_source=facebook.com&utm_campaign=buffer#section2.

102 Canada, Indigenous Services Canada, 2025, https://www.susanblacklin.com/s/Results-SFNWA-June-30-2025-A-2024-00409pdf-1.pdf.

103 The Saskatchewan First Nations Water Association (@thesfnwa), "We are so grateful to @SaskWater for sponsoring the keynote address at the 2025 SFNWA Conference by the inspiring

Cadmus Delorme." Instagram, October 10, 2025, https://www.instagram.com/p/DPoxYIkAb_k.

104 SaskWater, "Refreshing solutions. Superior quality. Healthy partnerships across the province." November 30, 2025, https://www.saskwater.com.

105 Canada, Indigenous Services Canada, 2025, https://www.susanblacklin.com/s/SFNWA-Emojis.pdf.

106 Meaghan Myles, Andrea Cherry, "Proposed Updates to the Protocol for Centralised Drinking Water Systems in First Nations Communities: A Facilitated Discussion with Indigenous Services Canada," Strategic Water Management (SWM) Directorate of Indigenous Services Canada, December 5, 2025, https://www.susanblacklin.com/s/Proposed-Updates-to-the-Protocol-for-Centralised-Drinking-Water-Systems-in-First-Nations-Communities.pdf.

107 Melissa Hotain, email message to author, July 3, 2025.

108 The Saskatchewan First Nations Water Association, "Board Member, James Cappo from Muscowpetung Saulteaux Nation #80, put together a team of water operators from File Hills Qu'Appelle Tribal Council in TREATY FOUR to compete in the Echo Lake Plywood Regatta," Facebook, July 7, 2025, https://www.facebook.com/SaskatchewanFirstNationsWaterAssociation/posts/board-member-james-cappo-from-muscowpetung-saulteaux-nation-80-put-together-a-te/1086303126958134.

109 Bridge City Warmth, "BCW Community Dinner & Distribution Event - Happening Today!" Facebook, September 23, 2025, https://www.facebook.com/61571356512888/posts/bcw-community-dinner-distribution-event-happening-todayour-amazing-bcw-volunteer/122140353608711883.

110 Water First, "About Us," November 6, 2025, https://waterfirst.ngo/about-us.

111 Canada, Indigenous Services Canada, 2025, https://www.susanblacklin.com/s/Water-first-Funding_000.pdf.

112 Canada Revenue Agency, "Water First," November 15, 2025, https://apps.cra-arc.gc.ca/ebci/hacc/srch/pub/dsplyRprtngPrd?q.

srchNmFltr=Water%2BFirst&q.stts=0007&selectedCharityBn= 838525269RR0001&dsrdPg=1.

113 Water First, "Audited Financial Statements for the Year Ended August 31, 2023," December 5, 2025, https://www.susanblacklin.com/ s/Water-First-Audited-Financial-Statements-2022-2023.pdf; Water First, "Audited Financial Statements: Year Ended August 31, 2022," December 5, 2025, https://www.susanblacklin.com/s/Water-First-2021- 2022-Audited-Financial-Statements.pdf; Powell Jones LLP, "Water First Education and Training Inc. Financial Statements, August 31, 2021," December 5, 2025, https://www.susanblacklin.com/s/Water- First-2020-2021-Audited-Financial-Statements.pdf.

114 Canada, Indigenous Services Canada, 2025, https://www. susanblacklin.com/s/Water-First-00411-Consult-Package.pdf.

115 Scott Anderson, Andrew Culbert, Ionna Roumeliotis, "Canadian charities complicit in helping fund Israeli settlement movement in West Bank, critics say," CBC News, October 17, 2025, https://www.cbc.ca/news/canada/canadian-charities-israeli-settlement- movement-west-bank-9.6939381.

116 Leanne Sanders, "Findings of FSIN audit 'hurt' public opinion of First Nations governments says Indigenous political commentator," APTN News, September 25, 2025, https://www.aptnnews.ca/national- news/findings-of-fsin-audit-hurt-public-opinion-of-first-nations- governments-says-indigenous-political-commentator.

117 Canada, Indigenous Services Canada, 2025, https://www. susanblacklin.com/s/John-Millar-Minda-Richardson.pdf.

118 Charity Intelligence Canada, "Top 10 Impact Charities," November 15, 2025, https://www.charityintelligence.ca/charity- profiles/top-10-impact-charities.

119 Imagine Canada, "Imagine Canada," November 12, 2025, https://imaginecanada.ca/en.

120 Chantal Edwards, "Imagine Canada Hosts Panel on Ethical AI and the Opportunities & Challenges in the Nonprofit Sector," Imagine Canada, August 6, 2024, https://imaginecanada.ca/en/360/panel-on- ethical-ai-and-the-opportunities-challenges-nonprofit-sector.

121 Prime Minister of Canada, "Securing Canada's AI advantage,"

April 7, 2024, https://www.pm.gc.ca/en/news/news-releases/2024/04/07/securing-canadas-ai.

122 Ethan Phillips, "Solomon launches 30-day AI Task Force," Oversight, October 3, 2025, https://ethanphillips.substack.com/p/solomon-launches-30-day-ai-task-force.

123 Catharine Tunney, "Canada now has a minister of artificial intelligence. What will he do?" CBC News, May 17, 2025, https://www.cbc.ca/news/politics/artificial-intelligence-evan-solomon-1.7536218.

124 Water First, "Nobel Winner Donates Half of Prize Money to Water First," November 4, 2024, https://waterfirst.ngo/press-release/nobel-winner-donates-half-of-prize-money-to-water-first.

125 Sofia Gomez Tamayo, Andrea Petrelli, "Work Transformed: The Promise and Peril of Artificial Intelligence," International Labour Organization, June 2025, https://www.ilo.org/sites/default/files/2025-07/ilo%20brief%20work%20transformed%20promise%20and%20peril%20of%20ai.pdf.

126 The Associated Press, "Amazon cutting 14,000 corporate jobs to spend more on AI," CBC News, October 28, 2025, https://www.cbc.ca/news/business/amazon-14000-corporate-layoffs-artificial-intelligence-9.6956357.

127 Samantha Cole, "Judges Are Fed up With Lawyers Using AI That Hallucinate Court Cases," Court Watch, March 3, 2025, https://www.courtwatch.news/p/judges-are-fed-up-with-lawyers-using-ai-that-hallucinate-court-cases.

128 Water First, "Nobel Winner Donates Half."

129 Angus, *Dangerous memory: Coming of age in the decade of greed.*

130 Martin Coulter, "AI pioneer says its threat to world may be 'more urgent' than climate change," Reuters, May 9, 2023, https://www.reuters.com/technology/ai-pioneer-says-its-threat-world-may-be-more-urgent-than-climate-change-2023-05-05.

131 R/DailyShow (@ElectricGremlin), "Why Jon Stewart Called Shark Tank's Kevin O'Leary An 'A**hole' On The Daily Show," Reddit, November 17, 2025, https://www.reddit.com/r/DailyShow/

comments/1bquskg/why_jon_stewart_called_shark_
tanks_kevin_oleary.

132 Christian Roethling, "Kevin O'Leary is a traitor to Canada," The Brock Press, January 9, 2025, https://brockpress.com/kevin-oleary-is-a-traitor-to-canada.

133 Stephen Starr, "Are data centres a threat to the Great Lakes?" The Narwhal, May 20, 2025, https://thenarwhal.ca/great-lakes-data-centres-threat.

134 Reuters, "Power-Hungry," Watershed Sentinel, June 5, 2025, https://www.susanblacklin.com/s/Power-Hungry-Water-Sentinel-Summer-2025.pdf.

135 Jonathan Montpetit, Yvette Brend, "AI-related data centres use vast amounts of water. But gauging how much is a murky business," CBC News, October 22, 2025, https://www.cbc.ca/news/ai-data-centre-canada-water-use-9.6939684.

136 Emily Fagan, "Areas hard hit by B.C. drought now the target of bottled water corporations," The Narwhal, November 17, 2021, https://thenarwhal.ca/bc-drought-water-bottling.

137 Jen Groundwater, "Merville Water Guardians," Watershed Sentinel, November 29, 2022, https://watershedsentinel.ca/article/water-guardians.

138 Kate Bueckert, "Why Nestlé's Aberfoyle well matters so much to Guelph, Ont., residents," CBC News, September 26, 2016, https://www.cbc.ca/news/canada/kitchener-waterloo/nestle-guelph-rally-nestle-water-aberfoyle-1.3779649.

139 Tom Perkins, "The fights to stop Nestlé from taking America's water to sell in plastic bottles," The Guardian, October 29, 2019, https://www.theguardian.com/environment/2019/oct/29/the-fight-over-water-how-nestle-dries-up-us-creeks-to-sell-water-in-plastic-bottles.

140 Financial Statements 2018, Nestle Group. https://www.nestle.com/sites/default/files/asset-library/documents/library/documents/financial_statements/2018-financial-statements-en.pdf.

141 Wilson, Janet. "Nestlé is still taking national forest water for its Arrowhead label, with feds' help." *Deseret Sun.* June 13, 2019.

142 Conservation Ontario, "Walkerton Inquiry," November 16,

2025, https://conservationontario.ca/conservation-authorities/source-water-protection/history.

143 Advisories for Ontario, Water Today website. https://www.watertoday.ca/textm-p.asp?province=8

144 Water Canada, "Manitoba government providing financial support to test well water," May 5, 2025, https://www.watercanada.net/manitoba-government-providing-financial-support-to-test-well-water.

145 Safe Drinking Water Foundation, "Dr. Colin Fricker - Is The Indicator Dead? Microbial Monitoring of Water Supplies," YouTube Video, 1:16:51, November 16, 2025, https://www.youtube.com/watch?v=n5sfpDnqWPk&list=PLqB9O7SI3dp1Eu CVZKFm7qg3Pe2uu3nCz&index=2.

146 Water Canada, "Manitoba government providing financial support."

147 Ecojustice, "Tell the federal government to keep mandatory checks on pesticide safety," November 30, 2025, https://ecojustice.ca/take-action/tell-the-federal-government-to-keep-mandatory-checks-on-pesticide-safety.

148 Annabelle Olivier, "Quebec study finds 'cocktail of different pesticides' in treated tap water," CBC News, May 13, 2025, https://www.cbc.ca/news/canada/montreal/treated-drinking-water-pesticides-quebec-chateauguay-river-1.7532362.

149 Olivier, "Quebec study finds."

150 LégisQuébec, "Regulation respecting the quality of drinking water," November 16, 2025, https://www.legisquebec.gouv.qc.ca/fr/document/rc/q-2,%20r.%2040?langCont=en.

151 Olivier, "Quebec study finds."

152 Olivier, "Quebec study finds."

153 Olivier, "Quebec study finds.

154 Institut national de la recherche scientifique, "Maryse Bouchard," June 2, 2025, https://inrs.ca/en/research/professors/maryse-bouchard.

155 Xiameng Feng, Zhen Liu, Sung Vo Duy, Lise Parent, Benoit Barbeau, Sébastien Sauvé, "Temporal trends of 46 pesticides and 8 transformation products in surface and drinking water in Québec,

Canada (2021–2023): Potential higher health risks of transformation products than parent pesticides," *Water Research*, vol. 277, 2025: 123339, https://doi.org/10.1016/j.watres.2025.123339.

156 Olivier, "Quebec study finds."

157 Indigenous Services Canada, "Ending long-term drinking water advisories," November 18, 2025, https://www.sac-isc.gc.ca/eng/1506514143353/1533317130660.

158 British Columbia Ministry of Health Planning and Ministry of Health Services, "Action Plan for Safe Drinking Water in British Columbia," Government of British Columbia, December 5, 2025, https://www2.gov.bc.ca/assets/gov/environment/air-land-water/safe_drinking_printcopy.pdf.

159 The Conference Board of Canada, "Water Quality Index," June 27, 2023, https://www.conferenceboard.ca/hcp/water-quality-index-aspx.

160 Hannah Moffatt, Sylvia Struck, "Water-borne Disease Outbreaks in Canadian Small Drinking Water Systems," National Collaborating Centres for Public Health, November 2011, https://ncceh.ca/sites/default/files/SDWS_Water-borne_EN.pdf.

161 The Conference Board of Canada, "Water Quality Index."

162 Xiameng Feng, Zhen Liu, Sung Vo Duy, Lise Parent, Benoit Barbeau, Sébastien Sauvé, "Temporal trends of 46 pesticides."

163 Delores Broten, "Tracking Fracking: Communities need to know gas extraction chemicals," *Watershed Sentinel*, July 11, 2025, https://www.susanblacklin.com/s/Watershed-Sentinel-Tracking-Fracking-1.pdf.

164 Martin LaSalle, "How Sébastien Sauvé turned 'forever chemicals' into big news," UdeMNouvelles, April 15, 2025, https://nouvelles.umontreal.ca/en/article/2025/04/15/how-sebastien-sauve-turned-forever-chemicals-into-big-news.

165 Leah Borts-Kuperman, "Employees on a Canadian military base say contamination is making them sick. Here's what you need to know," The Narwhal, July 16, 2025, https://thenarwhal.ca/canadian-armed-forces-moose-jaw-explainer.

166 Suzanne Fenton, Alan Ducatman, Alan Boobis, Jamie DeWitt,

Christopher Lau, Carla Ng, James Smith, Stephen Roberts, "Per- and Polyfluoroalkyl Substance Toxicity and Human Health Review: Current State of Knowledge and Strategies for Informing Future Research," *Environmental Toxicology and Chemistry*, vol. 40, no. 3, March 2021, https://pubmed.ncbi.nlm.nih.gov/33017053.

167 National Academies of Sciences, Engineering, and Medicine; Health and Medicine Division; Division on Earth and Life Studies; Board on Population Health and Public Health Practice; Board on Environmental Studies and Toxicology; Committee on the Guidance on PFAS Testing and Health Outcomes, "Guidance on PFAS Exposure, Testing, and Clinical Follow-Up," *National Academies Press*, 2022: 35939564, https://pubmed.ncbi.nlm.nih.gov/35939564.

168 Water Canada, "Experts gather in Toronto to tackle PFAS management in Canada," September 18, 2025, https://www.watercanada.net/experts-gather-in-toronto-to-tackle-pfas-management-in-canada.

169 Broten, "Tracking Fracking."

170 Phoebe Weston, "'Even if we stop drinking we will be exposed': a French region has banned tap water. Is it a warning for the rest of Europe?" The Guardian, July 1, 2025, https://www.theguardian.com/environment/2025/jul/01/pfas-forever-chemicals-water-contamination-saint-louis-france-aoe.

171 Samantha Beattie, "Line 5 pipeline between US and Canada could cause 'devastating damage' to Great Lakes, say environmentalists," CBC News, August 3, 2021, https://www.cbc.ca/news/canada/toronto/line-five-environment-great-lakes-1.6120882.

172 Todd Richmond, "US army engineers fast-track Great Lakes tunnel permits under Trump emergency order," CBC News, April 16, 2025, https://www.cbc.ca/news/canada/windsor/army-fast-track-great-lakes-tunnel-permits-1.7511851.

173 Ian Bickis, "Enbridge says federal support for oil and gas 'remains to be seen'," CBC News, August 1, 2025, https://www.cbc.ca/news/canada/calgary/enbridge-energy-canada-federal-1.7599825.

174 Water Canada, "Federal boost to tackle Great Lakes pollution and secure clean water for communities," September 20, 2024, https://

www.watercanada.net/federal-investment-to-clean-great-lakes-pollution-and-ensure-clean-water-for-communities.

175 MDA Space, "MDA to Build 17 Satellites to Enhance Globalstar's Leo Constellation," February 24, 2022, https://mda.space/article/mda-to-build-17-satellites-to-enhance-globalstars.

176 Water Canada, "Federal boost to tackle."

177 Chiefs of Ontario, "Chiefs of Ontario issue statement on cancellation of herbicide spraying in the North Shore area," August 28, 2025, https://chiefs-of-ontario.org/chiefs-of-ontario-issue-statement-on-cancellation-of-herbicide-spraying-in-the-north-shore-area.

178 CBC News, "Toxic algal blooms threaten Canadian drinking water, but scientists are developing a way to protect against the harmful slime," October 16, 2024, https://newsinteractives.cbc.ca/features/2024/toxic-algae.

179 Thomson Reuters, "South Australia algal bloom a 'natural disaster,' state's premier says, as species wiped out," CBC News, July 22, 2025, https://www.cbc.ca/news/climate/south-australia-algal-bloom-disaster-1.7590816.

180 Government of Saskatchewan, "Province Invests into SRC to Continue Innovative Research and Technology Work," March 20, 2024, https://www.saskatchewan.ca/government/news-and-media/2024/march/20/province-invests-into-src-to-continue-innovative-research-and-technology-work.

181 Government of Saskatchewan, "Province Invests into SRC."

182 Naira Hofmeister, Crispin Dowler, Laurent Gaberell, "'If it's dangerous for one population, it will be for the other': the Brazilian farmers poisoned by a banned pesticide exported from Britain," Unearthed, August 12, 2024, https://unearthed.greenpeace.org/2024/12/08/brazil-farmers-poisoned-banned-pesticide-diquat-syngenta-export-uk.

183 Hofmeister, Dowler, Gaberell, "'If it's dangerous for one population."

184 Simon Marks, Giulia Paravicini, "How Syngenta won the war over weedkillers," Politico, March 19, 2018, https://www.politico.eu/

article/how-syngenta-swiss-agrichemical-avoided-weedkiller-pesticide-ban-despite-safety-concerns-eu-commission.

185 Pesticide Action Network Europe, "German TV documentary on glyphosate, with PAN-Europe Hans Muilerman interview," May 3, 2024, https://www.pan-europe.info/node/2700.

186 J.W. Hodgeson, W.J. Bashe, J.W. Eichelberger, J.W. Munch, "Method 549.2: Determination of Diquat and Paraquat in Drinking Water by Liquid-Solid Extraction and High Performance Liquid Chromatography with Ultraviolet Detection," National Exposure Research Laboratory Office of Research and Development US Environmental Protection Agency, June 1997, https://www.epa.gov/sites/default/files/2015-06/documents/epa-549.2.pdf.

187 B.C. Pearce, O.E. Langer, "Impact of Aquatic Herbicides (Diquat-Paraquat) on Selected Organisms in Okanagan Lake, British Columbia," Fisheries and Environment Canada, Water Quality Division, Habitat Protection Directorate, Pacific Region, June 1977, https://waves-vagues.dfo-mpo.gc.ca/Library/142524.pdf.

188 Lucy Sharratt, "Mad Dog Down: Health Canada ordered to reassess glyphosate risks," *Watershed Sentinel*, April 29, 2025, https://www.susanblacklin.com/s/Watershe-Sentinal-p7-spring-2025-Mad-Dog-Down-1.pdf.

189 Sharratt, "Mad Dog Down."

190 Food Tank, Beveridge & Diamond PC, "Hell No, GMO!" *Watershed Sentinel*, Summer 2025, https://www.susanblacklin.com/s/Watershed-Sentinel-Hell-No-GMO-1.pdf.

191 Kristi Heim, Maureen O'Hagan, "Local activists are mounting a campaign to get the Gates Foundation to cut its ties with the agribusiness giant Monsanto and other firms involved in developing bioengineered crops." The Seattle Times, August 28, 2010, https://www.seattletimes.com/seattle-news/gates-foundation-ties-with-monsanto-under-fire-from-activists.

192 Gibson, Kate. "Pesticide linked to reproductive issues found in Cheerios, Quaker Oats and other oat-based foods." CBS News online. February 20, 2024. https://www.cbsnews.com/news/cheerios-quaker-oats-infertility-chemicals-in-cereal-ewg/

193 Brian Zinchuk, "Federal $2B contribution to Ontario nuclear project has major implications for Saskatchewan," Your West Central Voice and Chronicle, October 28, 2025, https://www.yourwestcentral.com/articles/federal-2b-contribution-to-ontario-nuclear-project-has-major-implications-for-saskatchewan.

194 No Nuclear Waste In Northwestern Ontario, "NNWNO No Nuclear Waste in Northwestern Ontario. We are a task force taking action to stop the burial of high-level nuclear waste in Northwestern Ontario." Facebook, November 16, 2025, https://www.facebook.com/NoNuclearWasteInNWO.

195 Mandell Pinder, "Kebaowek First Nation v Canadian Nuclear Laboratories 2025, FC 319 - Case Summary," Mandell Pinder LLP, May 14, 2025, https://www.mandellpinder.com/kebaowek-first-nation-v-canadian-nuclear-laboratories-2025-fc-319-case-summary.

196 Canadian Nuclear Laboratories, "The science of tomorrow, today." August 22, 2025, https://www.cnl.ca.

197 APTN News, "Kebaowek First Nation holds news conference in Ottawa over Chalk River proposal," YouTube Video, 1:19:44, November 16, 2025, https://www.youtube.com/watch?v=6dRCB0j-Sro.

198 Fasken Martineau DuMoulin LLP, "Another Sign of Canadian Leadership in Nuclear Energy: North America's First SMR Has Received Federal and Provincial Approval," June 4, 2025, https://www.fasken.com/en/knowledge/2025/06/another-sign-of-canadian-leadership-in-nuclear-energy-north-americas-first-smr-has-received-approval.

199 Stefan Labbé, "Canada to ban open net fish farms in B.C. by 2029," Times Colonist, June 20, 2024, https://www.timescolonist.com/local-news/canada-to-ban-open-net-fish-farms-in-bc-by-2029-9105139.

200 Tom Latourrette, Fabian Villalobos, Elisa Yoshiara, Zohan Hasan Tariq, "The Potential Impact of Seabed Mining on Critical Mineral Supply Chains and Global Geopolitics," RAND, April 9, 2025, https://www.rand.org/content/dam/rand/pubs/research_reports/RRA3500/RRA3560-1/RAND_RRA3560-1.pdf

201 Ofwat, "Improving life through water," November 16, 2025, https://www.ofwat.gov.uk.

202 The Associated Press, "Britain's biggest water firm hit with record fine over sewage and dividends," BNN Bloomberg, May 28, 2025, https://www.bnnbloomberg.ca/business/company-news/2025/05/28/britains-biggest-water-firm-hit-with-record-fine-over-sewage-and-dividends.

203 The Associated Press, "Britain's biggest water firm."

204 AJ Bell, "Thames Water pays £158 million dividend amid scramble for funding," July 9, 2024, https://www.ajbell.co.uk/news/articles/thames-water-pays-ps158-million-dividend-amid-scramble-funding.

205 AJ Bell, "Thames water pays."

206 Windrush Against Sewage Pollution, "Sewage pollution is killing our rivers," November 16, 2025, https://www.windrushwasp.org/sewage-pollution.

207 Sarah Wolfe, Mirae Kim, Yu Xia, "Water shortages in Canada? Drought and rising demand threaten Metro Vancouver's water supply despite conservation efforts," Water Canada, September/October 2025, https://emagazine.watercanada.net/?pid=ODk8907897&p=35&v=1.8.

208 Julia-Simone Rutgers, "'Afraid of the water'? Life in a city that dumps billions of litres of raw sewage into lakes and rivers," The Narwhal, May 10, 2025, https://thenarwhal.ca/winnipeg-sewage-leak-overflows.

209 Julia-Simone Rutgers, "From muddy to cruddy: When it comes to the human-waste sullied waters of the Red River, it's a matter of look, don't touch," Winnipeg Free Press, May 9, 2025, https://www.winnipegfreepress.com/featured/2025/05/09/from-muddy-to-cruddy.

210 Julia-Simone Rutgers, "From muddy to cruddy."

211 Julia-Simone Rutgers, "'Afraid of the water'?"

212 Julia-Simone Rutgers, "5 things to know about Winnipeg's big sewage problem," The Narwhal, May 13, 2025, https://thenarwhal.ca/winnipeg-sewage-leak-numbers.

213 Government of Canada, "Federal and provincial governments invest in sewer system transformation for Oak Bay," February 15,

2024, https://www.canada.ca/en/housing-infrastructure-communities/news/2024/02/federal-and-provincial-governments-invest-in-sewer-system-transformation-for-oak-bay.html.

214 City of Toronto, "Don River and Central Waterfront & Connected Projects," November 30, 2025, https://www.toronto.ca/services-payments/water-environment/managing-rain-melted-snow/what-the-city-is-doing-stormwater-management-projects/lower-don-river-taylor-massey-creek-and-inner-harbour-program.

215 AC Minerals Group, "Affordable, scalable, profitable & circular," Asbeter, November 16, 2025, https://www.asbeter.com/en/breakthrough.

216 American Cancer Society, "Asbestos and Cancer Risk," November 16, 2025, https://www.cancer.org/cancer/risk-prevention/chemicals/asbestos.html.

217 Nancy Carswell, "Dangers of Asbestos Cement Water Pipes," Saskatchewan Green Party, April 17, 2023, https://www.saskgreen.ca/asbestos_water_pipes.

218 Jock McCulloch, Geoffrey Tweedale, "Shooting the messenger: the vilification of Irving J. Selikoff," International Journal of Health Services, vol. 37, no. 4, 2007, https://pubmed.ncbi.nlm.nih.gov/18072311.

219 Oxfam Canada, "Why Extreme Inequality."

220 Anirudh Bhattacharyya, "Canada: Petition calling for no confidence vote against Trudeau govt garners 300K signatures," Hindustan Times, December 14, 2023, https://www.hindustantimes.com/world-news/canada-petition-calling-for-no-confidence-vote-against-trudeau-govt-garners-300k-signatures-101702531216297.html.

221 Christian Paas-Lang, "A record number of people signed e-petitions last year - do they make a difference?" CBC News, January 6, 2024, https://www.cbc.ca/news/politics/epetitions-2023-increasing-popularity-1.7074358.

222 ch.ch, "The referendum," November 16, 2025, https://www.ch.ch/en/votes-and-elections/referendum.

223 Rachel Gilmore, "These were the most-signed e-petitions sent to the Trudeau government," CTV News, August 1, 2019, https://www.

ctvnews.ca/politics/article/these-were-the-most-signed-e-petitions-sent-to-the-trudeau-government.

224 May, *Losing Confidence: Power, Politics, and the Crisis in Canadian Democracy.*

225 Dr. Hans Peterson, "Biological Water Treatment Discussed at UN," *Canadian Water Treatment*, May/June 2005, https://www.susanblacklin.com/s/20-CWT-Biological-Water-Treatment-Discussed-at-UN.pdf.

226 Carrington, Damian. "Era of 'Global Water Bankruptcy' is Here: UN Report Says." The Guardian. January 20, 2026. https://www.theguardian.com/environment/2026/jan/20/era-of-global-water-bankruptcy-is-here-un-report-says?CMP=share_btn_url

227 "Minister not promising source water protection in new First Nations clean water bill." CTV News online. December 30, 2025.https://www.ctvnews.ca/canada/article/minister-not-promising-source-water-protection-in-new-first-nations-clean-water-bill/

228 Invisible Ingredients: Tackling Toxic Chemicals in the Food System." Systemiq, 2025.

229 Canadians and Canadian Organizations That Have Endorsed This Petition https://static1.squarespace.com/static/688d75c0c0336d73f60b3f58/t/690a74b5bdf345277b3a962c/1762292917560/Canadians+Need+Legally+Enforceable+Regulations+for+their+Drinking+Water%21+%289%29.pdf.

About the Author

Susan Blacklin wrote and published her first memoir in 2024 at age 73. *Water Justice* is Susan's second book, a call to action. Susan grew up in London, England, and fulfilled her dream of emigrating to Canada at eighteen years of age.

For fifteen years she and her late ex-husband, Dr. Hans Peterson, founded and ran the Safe Drinking Water Foundation, volunteering tirelessly to help First Nations communities in their quest for truly safe drinking water.

Susan has lived in Winnipeg, Manitoba as well as Saskatoon and Outlook, Saskatchewan. Now retired, she resides in Qualicum Beach on Vancouver Island, British Columbia. Every day she appreciates that she now lives an idyllic life with her partner Roelof, and when she isn't writing she enjoys many hobbies, including gardening, tutoring, painting, knitting, and walking the local beach in the summertime.

Please visit www.susanblacklin.com to subscribe to Susan's newsletter.

Endorsements

"The asbestos pipe risks that Susan reveals are worth nightmares of their own and the complexities of governance, responsibility and regulation, or rather the failure of all of these to ring a very loud bell to wake up a population heading towards a cliff edge, brings a chilling message - avoiding this disaster is probably going to be down to the victims; the public who need to start making a fuss and a big one."

Ash Smith, Founder, Windrush Against Sewage Pollution (UK)

Water Justice: What You Don't Know Could Kill You exposes threats to the health of Canadians from Canada's drinking water, revealing government negligence affecting both Indigenous and non-Indigenous communities. The book critiques Canada's weak and provincially varied drinking water guidelines compared to other G7 nations' drinking water regulations, arguing that corporate lobbyists hold more influence than citizens over water policy.

Blacklin documents political indifference across parties and connects the decline of Canada's water quality to the erosion of our democracy. Highlighting contaminated source waters and inadequate protections, the book calls readers to action, emphasizing that safe drinking water requires citizen engagement to overcome governmental failure and corporate control.

"I have come to know Susan through her writing and advocacy work. She is a passionate Canadian, who cares deeply about the water we all rely on. She is a well-researched, and compelling storyteller, who deserves to be heard. Canada needs more Susan Blacklin's."

Julian Branch, Board Member, Prevent Cancer Now

Manufactured by Amazon.ca
Bolton, ON

54980028R00234